Bilkent University Lecture Series

Bilkent University Lecture Series

Rolf Färe   Shawna Grosskopf

Cost and Revenue Constrained Production

With 33 Figures

Springer-Verlag
New York Berlin Heidelberg London Paris
Tokyo Hong Kong Barcelona Budapest

Prof. ROLF FÄRE

Department of Economics
College of Liberal Arts
Southern Illinois University
  at Carbondale
Carbondale, IL 62901
USA

Prof. SHAWNA GROSSKOPF

Department of Economics
College of Liberal Arts
Southern Illinois University
  at Carbondale
Carbondale, IL 62901
USA

Bilkent University Lecture Series
Mathematical Economics and
  Finance Track Editor

Advisory Board Chairman

Prof. Sübidey Togan
Faculty of Economics, Administrative and
  Social Sciences
Bilkent University
Bilkent, 06533 Ankara, Turkey

Prof. Ali Doğramacı
Office of the Board of Trustees
Bilkent University
Bilkent, 06533 Ankara, Turkey

Library of Congress Cataloging-in-Publication Data
Färe, Rolf, 1942–
    Cost and revenue constrained production / Rolf Färe, Shawna
Grosskopf.
        p.   cm. – (Bilkent University lecture series ; v. 4)
    Includes bibliographical references and index.
    ISBN-13: 978-1-4612-7613-5   e-ISBN-13: 978-1-4612-2626-0
    DOI: 10.1007/978-1-4612-2626-0

    1. Production functions (Economic theory).   2. Econometric models.
3. Duality theory (Mathematics).   I. Grosskopf, Shawna.   II. Title.
III. Series.
HB241.F326   1994
338.5 – dc20   93-32408

Printed on acid-free paper.

Camera-ready copy prepared by the authors.

9 8 7 6 5 4 3 2 1

Preface

The basic notion underlying this monograph – budget or revenue constrained models of production – we owe to Ronald W. Shephard, who recognized its fundamental importance in modeling behavior in a wide variety of settings including the service and public sector. Our endeavor here is to extend Shephard's earlier work in several directions while maintaining his axiomatic approach.

Our contributions include an expanded set of duality results and a general bent toward empirical implementation: including various parameterizations, applications to efficiency and productivity measurement, and shadow pricing. We hope to provide those engaged in empirical work with some powerful and useful tools which have received relatively little attention.

The nature of the material in this monograph is somewhat technical, however, the level of mathematical difficulty is standard. Although we have tried to keep the monograph fairly self-contained, we have also kept technical detail to a minimum in the body of the text. Many technical extensions appear as problems at the ends of Chapters. The reader is also referred to the notes at the end of each chapter for references to additional literature.

A prepublication draft of this manuscript was used as lecture notes in a graduate course in production theory at the Department of Economics at Bilkent University. We thank our students as well as faculty members for their patience and interest. Special thanks go to Dean Togan, Zeynap Köksal and Ali Dogramaci for making our stay in Ankara not only productive, but also enjoyable.

Knox Lovell and Wolfgang Eichhorn read earlier versions of this manuscript. We have integrated their extensive comments and suggestions in the current version. We are extremely grateful for their careful, constructive suggestions, which have done much to enhance this work.

This manuscript was produced by Mariann Baratta who transformed it from handwritten notes, through revisions to the document you see here. We are grateful for her outstanding work.

Contents

Chapter 1

Introduction

1.0 Motivation

The traditional theory of the firm which serves as the underpinnings of what we know as neoclassical production theory, has proved to be a useful benchmark model. The implied behavioral goals of profit maximization, revenue maximization and cost minimization are useful in understanding choices made at the firm and industry level, as well as serving as quite powerful models for empirical analysis. Duality theory has enriched the use of cost, revenue and profit functions by providing a means of retrieving information on underlying technology from functions which are better suited to econometric estimation than production functions, for example. These alternative 'dual' representations of technology also provide those pursuing applied work with choices with respect to data requirements.

These basic optimization models have also served as points of departure; for example, there is a growing literature which seeks to identify the degree of deviation from the cost minimization, profit and revenue maximization standard. Related to that literature, attention is being paid to specifying the associated cost, revenue and profit functions as frontiers. Other examples of modifications of the standard paradigm include attempts to allow for price distortion due to regulation, utility maximizing decision-makers, etc. Incentive problems arising from principle-agent problems are also playing an important role in defining what is becoming known as the new theory of the firm.

The purpose of this monograph is to provide a class of models that builds on our basic models, yet generalizes them to model the case in which the decision-maker faces a budget constraint or revenue target. These models were termed indirect by Shephard, which illustrates the symmetric role played by indirect utility functions in demand theory. Our two major goals in pursuing these constrained or indirect models are: (1) to provide alternative behavioral goals which augment and generalize existing models, and (2) to provide those pursuing applied work with consistent models with alternative data requirements.

In pursuing our goals of providing alternative behavioral models and a menu of data choices, our basic approach is an axiomatic one, based on axioms which are as unrestrictive as possible while providing sufficient structure to be of practical use. In order to provide alternative data choices and to relate these models to existing models, we rely heavily on duality theory. Another feature of this work is the reliance on (direct and indirect) Shephard-type distance functions as representations of technology. These have several advantages over the more traditional production functions. First of all, like production functions, they are scalar valued functions, unlike production functions they readily model multiple output, multiple input technology. Second, distance functions have natural dual relationships to value or support functions. Although rarely exploited, distance functions are easy to estimate – in fact they are reciprocal to the widely used Farrell efficiency measures and the Debreu coefficient of resource utilization. As such they prove useful as performance measures, which we exploit here. Since we hope that this work will prove useful in applied work, we also provide parametric and nonparametric alternatives for calculating distance functions as well as the other indirect functions introduced in this monograph.

Our indirect models can be summarized into two broad categories: (1) those in which the decision-maker optimizes in the presence of a budget constraint or cost target, and (2) those in which the decision-maker faces a minimal revenue target. These restrictions can be used to augment existing models of output maximization, revenue maximization, input minimization, cost minimization, and profit maximization. These constrained models also provide information on underlying technology through duality and also add to our stock of dual relationships.

These indirect models provide what we believe to be useful characterizations of behavior. For example, we would argue that what we call the cost indirect output distance function is a useful model for the public sector: it assumes that the decision-maker has a fixed budget, and must choose inputs which satisfy that budget, while providing the maximum feasible services from those budget restricted resources. Put differently, the cost indirect output distance function seeks to maximize multiple outputs subject to a budget constraint. In contrast to the direct distance

2

function, which takes inputs as given (and therefore not subject to choice), and seeks to maximize outputs, the indirect output distance function takes the budget rather than inputs as given, allowing the decision-maker options concerning the choice of those inputs. In addition, as a distance function it readily models multiple output technology and also yields measures of performance – of particular usefulness in public sector applications.

As an example of a model which incorporates a revenue target, consider the revenue indirect cost function. Here the goal is to minimize costs given a minimal revenue target. Imagine a department in a large firm or a franchise operation which must meet a revenue or sales target and is competing with other units to contain cost. This model is more appropriate in these cases than a simple cost minimization model. Note also that it has different data requirements: the cost minimization model requires data on cost, input prices and output quantities, the revenue indirect cost model requires data on cost, input prices, target revenue and output prices (rather than output quantities).

1.1   Outline

We begin with the basics in Chapter 2: specification of the parent direct and indirect technologies, as well as the maintained axioms which they satisfy. Given these axioms, we derive the properties satisfied by the parent technology and inherited by the indirect technologies. We also include characterizations of the cost and revenue functions which are used to define the indirect technologies.

In Chapter 3 we turn to function representations of direct and indirect (i.e., cost or revenue restricted) technology. Here we focus on the aforementioned distance functions, which fully characterize multiple output technology and tell us "how far" from the frontier of that technology any given observation lies. We introduce cost and revenue restricted distance functions which judge how far an observation lies from the frontier of the corresponding cost or revenue constrained technology. These will serve as basic tools in later chapters. Included are "flexible" parameterizations of various distance functions as well as piecewise linear representations which can be employed in a linear programming framework.

In Chapter 4 we turn to modeling the indirect or constrained versions of revenue maximization, cost minimization and profit maximization. Here we exploit the dual relationship between distance and value or support functions and derive a series of indirect Shephard-type lemmas. These will prove useful in Chapter 5 in uncovering shadow or virtual prices and quantities. Parameterized models are included.

Introduction of these indirect models introduces the possibility of extending the duality results of the direct cases to these cost and revenue constrained cases. In fact, four "new" dualities are introduced in Chapter 5. We also show how to derive shadow prices of inputs and outputs from our indirect models, including translog parameterizations as examples.

In Chapter 6 we exploit the relationship between distance functions and Farrell-type efficiency measures and introduce a series of indirect measures of efficiency. Farrell-type decompositions into allocative and technical components as well as scale efficiency are included. We also provide nonparametric and parametric models for calculation of these efficiency measures. The former employs the piecewise linear models introduced in Chapter 3.

The final chapter focuses on productivity, i.e., change in performance and technology over time. Malmquist productivity indexes are shown to be composed of a series of distance functions, and indirect versions of these productivity indexes are introduced. Since we have shown how to calculate distance functions in previous chapters, it is a simple matter to modify them to calculate direct and indirect Malmquist productivity indexes. Thus these indexes can be calculated, and, since they allow for inefficiency and minimal data requirements, they have several advantages over the more popular Törnqvist indexes. We show under what conditions these indexes coincide. We also introduce indirect variations of Fisher ideal indexes and show when they coincide with their direct counterparts and when they coincide with Malmquist indexes.

Appendix A contains standard notation and mathematical definitions.

Chapter 2

Structure of the Direct and Cost and Revenue Restricted Technology

2.0 Introduction

This chapter describes the structure of the underlying unrestricted technology (also referred to here as the direct or parent technology) as well as the cost and revenue restricted technologies which are the focus of this study.

We model the parent technology in three equivalent ways: (1) as an input correspondence which specifies the subset of input vectors capable of producing a given output vector, (2) as an output correspondence which specifies the subset of output vectors obtainable from a given input vector, and (3) as the graph which is defined in terms of the input and output correspondences. Although equivalent, they highlight different aspects of production. The input (output) correspondence readily illustrates input (output) substitutability, whereas the graph best illustrates scale properties of technology.

We take an axiomatic approach to imposing structure on the parent technology, which in turn imposes structure on the restricted technologies. Section 2.1 of this chapter introduces the parent technology, i.e., the input correspondence, output correspondence and the graph as well as a set of axioms (including a subset of axioms to be maintained throughout this monograph) which provide structure without imposing unnecessary restrictions. These include assumptions concerning disposability of inputs and outputs, scaling properties, and feasibility among others.

Section 2.2 introduces the cost restricted technology—the cost indirect output correspondence—which specifies the subset of all output vectors obtainable from a prespecified

5

input price vector and budget (rather than a prespecified input vector).

We also briefly introduce the cost function and use it to provide alternative definitions of the cost indirect input correspondence. In addition, we derive the properties which the cost indirect input correspondence inherits from the set of maintained axioms imposed on the parent technology.

Section 2.3 focusses on a comparison of the (parent) input correspondence and the cost indirect technology.

Next we turn to the introduction of the revenue restricted technology—the revenue indirect input correspondence—which specifies the subset of input vectors capable of yielding target revenue given output prices (rather than a given output vector). We introduce the revenue function and define the revenue indirect technology with respect to it. We also derive the properties which the revenue indirect technology inherits from the parent technology when the set of maintained axioms is imposed.

The appendix includes further characterization of the cost and revenue functions and the properties they inherit from those maintained on the parent technology. These will prove useful later in the monograph.

A set of problems follows and notes on related literature concludes this chapter.

2.1   The Parent Technology

The parent or direct steady state (time independent) technology transforming input vectors (factors of production) $x = (x_1, \cdots, x_N) \in \Re_+^N$ into output vectors (goods produced) $u = (u_1, \cdots, u_M) \in \Re_+^M$ is modeled by the output correspondence P, the input correspondence L, or by the graph GR. Since we assume that input and output vectors are elements among nonnegative real numbers, divisibility is implied, but positivity of each component is not required, i.e., some zeros are allowed.

The *Output Correspondence*

$$(2.1.1) \qquad P : \Re_+^N \longrightarrow P(x) \subseteq \Re_+^M$$

maps input vectors $x \in \Re_+^N$ into subsets $P(x)$ of output vectors. The set $P(x)$ is termed the *Output Set* and it denotes the set of all output vectors $u \in \Re_+^M$ that can be produced in unit time

6

by the input vector $x \in \Re^N_+$.

The *Input Correspondence*

(2.1.2) $L : \Re^M_+ \longrightarrow L(u) \subseteq \Re^N_+$

maps output vectors $u \in \Re^M_+$ into subsets of input vectors $x \in \Re^N_+$. The *Input Set* $L(u)$ denotes all input vectors that can produce (in unit time) the output vector u, $u \in \Re^M_+$. The input and the output correspondences are inverses in the sense of

(2.1.3) $u \in P(x) \Longleftrightarrow x \in L(u)$,

i.e., u belongs to the output set for x if and only if x belongs to the input set for u. An input-output vector $(x, u) \in \Re^{N+M}_+$ is feasible if and only if $x \in L(u)$ or equivalently if and only if $u \in P(x)$.

The collection of all feasible input-output vectors define the *Graph*, i.e.,

$$
\begin{aligned}
(2.1.4) \quad GR &= \{(x, u) \in \Re^{N+M}_+ : u \in P(x), x \in \Re^N_+\} \\
&= \{(x, u) \in \Re^{N+M}_+ : x \in L(u), u \in \Re^M_+\}.
\end{aligned}
$$

The graph is thus derived from either the output or input correspondence, and conversely these two correspondences may be derived from the graph in accordance with

(2.1.5) $P(x) = \{u : (x, u) \in GR\}$ and $L(u) = \{x : (x, u) \in GR\}$.

Figure (2.1) illustrates the relationships among $P(x)$, $L(u)$ and GR. The graph is bounded by the x-axis and the ray (0a). Given the input x°, the corresponding output set $P(x^\circ)$ consists of the closed and bounded interval $[0, u^\circ]$ measured along the u-axis. The input set, i.e., the set of inputs that can produce u°, $L(u^\circ)$, consists of the closed, but not bounded, interval $[x^\circ, +\infty)$.

The above discussion is summarized in

(2.1.6)   **Proposition** : $u \in P(x) \Longleftrightarrow x \in L(u) \Longleftrightarrow (x, u) \in GR$.

Proposition (2.1.6) states that the input and output sets and the graph model the same production technology, although they highlight different aspects of it. The output set e.g., primarily models output substitution, in the sense of how one output may be traded off for some other output or outputs. This set is consequently often termed the transformation or production

7

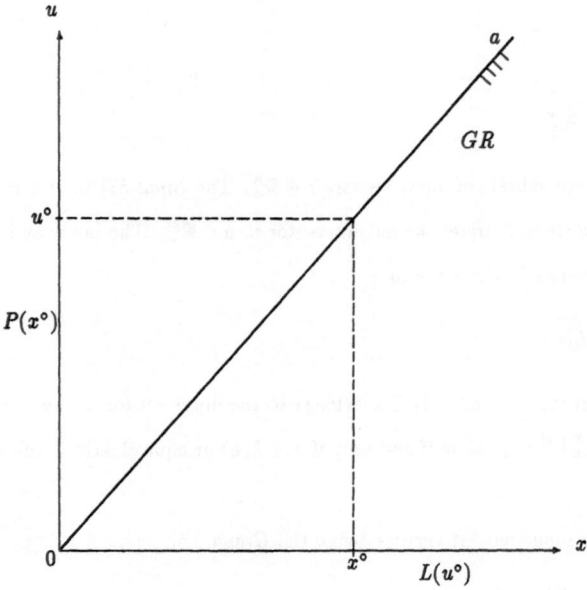

Figure 2.1: Representations of Technology

possibility set. On the other hand the input set treats output as given and models how one input may be substituted for some other input or inputs. Sometimes this set, reflecting the statement that x can produce u, is called the input requirement set. The graph, frequently referred to as the technology set, models input-output choices, and therefore can also be used to model profit maximization. The optimization problems defined on $P(x)$ and $L(u)$ are most often termed revenue maximization and cost minimization, respectively.

The production technology is assumed to satisfy certain axioms in order to be a valid model of production. These axioms should be commonly acceptable and should not impose unnecessary restrictions, yet they should provide enough structure to yield interesting results. The parent axioms considered here are only discussed for the output correspondence to avoid repetition. We begin with

P.1   (a) $0 \in P(x), \forall x \in \Re_+^N$;   (b) $u \notin P(0), u \geq 0$.

The first part of axiom P.1 states that the zero output vector belongs to the output set, for each input vector in the domain $\Re_+^N$. This means that inactivity or shutdown can occur as part of the

8

production process. Part (b) of the axiom models the nonexistence of a free lunch, i.e., input is required to produce output. Recall that $u \geq 0$ means that at least one component in the output vector is positive and none is negative.

As an exercise, we prove that (a) and (b) imply the equality $P(0) = \{0\}$. By (a), $\{0\} \subseteq P(0)$, thus assume that there exists an output vector $u \neq 0$, and $u \geqq 0$ such that $u \in P(0)$. This would imply that there is a $u \geq 0$ with $u \in P(0)$. However, in accordance with (b) this is not possible. This contradiction proves the claim.

Q.E.D.

P.2 $\forall x \in \Re_+^N, \lambda \geqq 1, P(x) \subseteq P(\lambda x).$

P.2.S $\forall x, y \in \Re_+^N, y \geqq x, P(x) \subseteq P(y).$

P.2 and P.2.S model disposability of inputs, or in mathematical terms monotonicity of the output correspondence. P.2 is referred to as weak disposability of inputs, and it states that if all inputs are increased in the same proportion i.e., $(\lambda x_1, \cdots, \lambda x_N)$, then no output is lost. This is in the sense that the original production set $P(x)$ is included in $P(\lambda x)$. Note, however, that $P(x)$ may equal $P(\lambda x)$, even when $\lambda > 1$. Strong disposability of inputs is modeled by P.2.S, and this disposability assumption states that if inputs are increased (or not decreased), the new output set contains the original. Clearly, if P.2.S holds, so does P.2. The converse however is not true. The two input disposability axioms are best illustrated in input space, thus we restate them in terms of the input set $L(u)$. P.2 and P.2.S are equivalent to

L.2 $\forall u \in \Re_+^M, x \in L(u)$ and $\lambda \geqq 1 \Longrightarrow \lambda x \in L(u),$

L.2.S $\forall u \in \Re_+^M, x \in L(u)$ and $y \geqq x \Longrightarrow y \in L(u),$

respectively.

Here we prove that P.2 $\Longleftrightarrow$ L.2. Thus assume that P.2 holds, and let $u \in P(x) \subseteq P(\lambda x)$, $\lambda \geqq 1$. By Proposition (2.1.6), $x \in L(u)$ and $\lambda x \in L(u)$, proving that P.2 $\Longrightarrow$ L.2. To show that L.2 $\Longrightarrow$ P.2, assume there exists a $u \in P(x)$ but $u \notin P(\lambda x)$ for some $\lambda \geqq 1$. Then by (2.1.6), $x \in L(u)$ but $\lambda x \notin L(u)$. This contradiction concludes the proof.

Q.E.D.

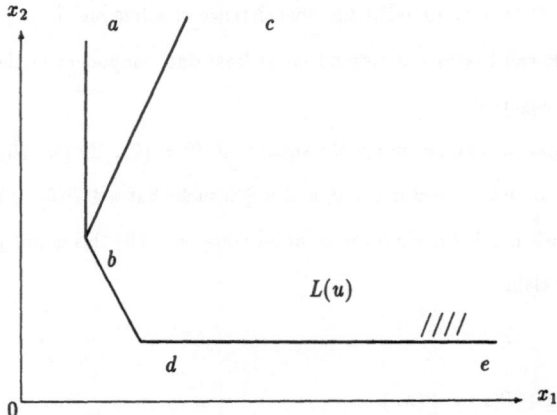

Figure 2.2: Disposability of Inputs

Next we turn to an illustration. Consider Figure 2.2. If inputs are only weakly disposable, i.e., P.2 applies but not P.2.S, a typical input set looks like $L(u)$, with $L(u)$ consisting of the area bounded by *(cbde)*. Weak disposability allows for "backward bending" isoquants like the segment *(c,b)*. Thus whenever one would like to model or allow for congestion (for example in transportation to allow for traffic congestion) one may assume that P.2 applies. However, one should not impose L.2.S, since strong disposability prevents "backward bending" isoquants. An input set in $\Re_+^2$ that satisfies L.2.S may look like the area bounded by *(abde)*.

P.3 $\quad \forall x \in \Re_+^N, u \in P(x)$ and $0 \leqq \theta \leqq 1 \Longrightarrow \theta u \in P(x)$.

P.3.S $\quad \forall x \in \Re_+^N, u \in P(x)$ and $0 \leqq v \leqq u \Longrightarrow v \in P(x)$.

Weak and strong disposability of output are modeled by P.3 and P.3.S, respectively. Weak disposability allows feasible outputs, i.e., $u \in P(x)$, to be proportionally contracted and remain feasible, i.e., $0 \leqq \theta \leqq 1$ and $\theta u \in P(x)$. Strong disposability allows any feasible output to be freely disposed of, i.e., $u \in P(x)$ and $v \leqq u$ imply that v is feasible. If one plans to model production processes that produce both desirable and undesirable outputs one should not impose P.3.S, since that axiom allows for free disposability. P.3, however, is clearly applicable under such conditions. The two axioms are illustrated in Figure 2.3. An output set that satisfies weak but not strong

10

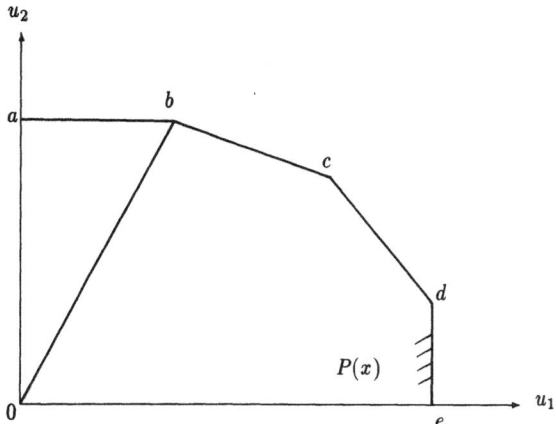

Figure 2.3: Disposability of Outputs

disposability may look like $P(x)$ bounded by $(0bcde0)$ while if strong disposability is satisfied the output set is enlarged to $(0abcde0)$.

P.4 $\forall x \in \Re_+^N$, $P(x)$ is bounded.

The boundedness of $P(x)$ is defined in terms of the norm on $\Re_+^M$ and means that there exists a sphere of finite radius that contains $P(x)$. In economic terms, boundedness models the property that one can only produce finite amounts of outputs from finite inputs.

P.5 P is a closed correspondence, i.e., if $x^\ell \longrightarrow x^\circ, u^\ell \longrightarrow u^\circ$ and $u^\ell \in P(x^\ell)$

   for all ℓ, then $u^\circ \in P(x^\circ)$.

The axiom that P is a closed correspondence is equivalent to the graph GR being a closed set. In addition, P.5 implies that for all $x \in \Re_+^N$, the output set $P(x)$ is closed, and thus the existence of a frontier of technology is ensured. P.5 then together with P.4 implies that the output sets are compact sets, i.e., closed and bounded in $\Re_+^M$.

P.6   If $u \in P(x), u \geq 0$, then $\forall \theta \geq 0$ there exists a scalar $\lambda \geq 0$ such that $\theta u \in P(\lambda x)$.

Axiom P.6 models attainability. It states that if a semipositive output vector $u \geq 0$ but $u \neq 0$ is producible, then any scalar extension of it can be produced by some scaling of the input vector in question.

11

Axioms P.1–P.6 are Shephard's basic axioms. An output correspondence P is called an *Output Production Correspondence* if and only if it satisfies P.1–P.6.

P.7 $\forall x \in \Re_+^N, P(x)$ is convex, i.e., $\forall u, v \in P(x), 0 \leqq \theta \leqq 1, (\theta u + (1 - \theta)v) \in P(x)$.

P.8 P is quasi $-$ concave, i.e., $\forall x, y \in \Re_+^N, 0 \leqq \lambda \leqq 1, P(x) \cap P(y) \subseteqq$

$P(\lambda x + (1 - \lambda)y)$.

P.9 P is concave, i.e., $\forall x, y \in \Re_+^N, 0 \leqq \lambda \leqq 1, \lambda P(x) + (1 - \lambda)P(y) \subseteqq$

$P(\lambda x + (1 - \lambda)y)$.

The three axioms P.7, P.8, and P.9 model various convexity properties. These are important in proving duality theorems. P.7 imposes convexity on $P(x)$, and P.8 imposes convexity of $L(u)$. To prove the latter claim, we need to show that P.8 implies that

L.8 $\forall u \in \Re_+^M, L(u)$ is convex.

Suppose that P.8 applies and that $u \in \Re_+^M$. If $u \notin P(x) \cap P(y)$ for any x and y in $\Re_+^N$, then $L(u)$ is empty and thus convex. Therefore assume that $u \in P(x) \cap P(y)$ for some x and y. By Proposition (2.1.6) and P.8, $x, y \in L(u)$ and for $0 \leq \lambda \leq 1, \lambda x \in (1 - \lambda)y \in L(u)$, proving our claim.

The last of the three convexity axioms, as we show next, implies that the graph is convex. Therefore, if (x, u) and $(y, v) \in GR$ we need to show that $(\lambda x + (1 - \lambda)y, \lambda u + (1 - \lambda)v) \in GR$. By Proposition (2.1.6), $u \in P(x)$ and $v \in P(y)$, thus P.9 implies that $\lambda u + (1 - \lambda)v \in P(\lambda x + (1 - \lambda)y)$ and again applying (2.1.6) yields, $(\lambda x + (1 - \lambda)y, \lambda u + (1 - \lambda)v) \in GR$. It is clear that neither P.7 or P.8 implies P.9, but that P.9 implies both P.7 and P.8.

The convexity of the graph, P.9, together with P.1 implies nonincreasing returns to scale in the following sense: The technology exhibits *Non Increasing Returns to Scale (NIRS)* if and only if $(x, u) \in GR$, and $0 < \mu \leqq 1 \Longrightarrow (\mu x, \mu u) \in GR$. To verify our claim we note from above that P.1(a) and (b) imply that $P(0) = \{0\}$. Let $u \in P(x)$ or equivalently $(x, u) \in GR$. By P.9, $\lambda P(0) + (1 - \lambda)P(x) \subseteqq P(\lambda 0 + (1 - \lambda)x)$, hence $(1 - \lambda)P(x) \subseteqq$ $P((1 - \lambda)x)$ and by Proposition 2.1.16, $((1 - \lambda)x, (1 - \lambda)u) \in GR$, concluding the proof.

12

Among the above axioms, the maintained subset imposed on the technology is (P.1–P.6), i.e., Shephard's basic axioms. Note that only weak disposability is assumed and that no convexity assumption is imposed.

Finally, define the sets $DomL = \{u \in \Re_+^M : L(u) \neq \emptyset\}$ and $DomP = \{x \in \Re_+^N : P(x) \neq \emptyset\}$. These sets are the *Effective Domains* of the input and output correspondences. $RangL = \{x \in \Re_+^N :$ there exists $u \in \Re_+^M$ such that $u \in P(x)\}$. Since P.1 applies, with $0 \in P(x), \forall x \in \Re_+^N, DomP = \Re_+^N$. Also note that $DomP = RangL$ and $DomL = RangP$, where $RangP = \{u \in \Re_+^M :$ there exists $x \in \Re_+^N$ such that $x \in L(u)\}$. The last two claims may be validated by Proposition (2.1.6).

2.2 The Cost Restricted Technology

The cost restricted technology, or as it is also called, the cost indirect technology is a mapping from cost deflated input prices into subsets of output vectors. More specifically, denote input prices by $p = (p_1, \cdots, p_N) \in \Re_+^N$ and the target cost or allowed expenditures by $C \in \Re_+$. The *Cost Indirect Output Correspondence* is a mapping

$$(2.2.1) \qquad IP : \Re_+^N \longrightarrow IP(p/C) \subseteq \Re_+^M$$

where

$$(2.2.2) \qquad IP(p/C) = \{u : u \in P(x), \sum_{n=1}^{N} p_n x_n \leqq C\} = \{u : u \in P(x), (p/C)x \leqq 1\}.$$

The *Cost Indirect Output Set* $IP(p/C)$ denotes all output vectors that are feasible, both technically, i.e., $u \in P(x)$, and economically, i.e., u can be produced at a cost not exceeding C. The cost C constrains the input vectors that can be used to the set $B(p/C) = \{x \in \Re_+^N : px \leqq C\}$. This set may be called the *Budget Set*. Some observations with regard to it are in order. The budget set is closed since $px \leqq C$. If all prices are positive, i.e., $p > 0$ or equivalently $p_n > 0, n = 1, \cdots, N$, then $B(p/C)$ is bounded for all $C \geqq 0$. Hence it is compact in $\Re_+^N$. To show the boundedness of $B(p/C)$, define $p_{n'} = \min_n \{p_n : n = 1, \cdots, N\}$, then the maximum that can be used of any input is $x_{n'} = C/p_{n'}$, if $C > 0$. This is a finite quantity, thus for $C > 0$, $B(p/C)$ is bounded. If $C = 0$, then each $x_n = 0, n = 1, \cdots, N$, and again $B(p/C)$ is bounded. The budget set includes the set of economically feasible input bundles, given C.

13

Prior to discussing the properties which the indirect output correspondence inherits from the parent technology we need to introduce the notion of a cost function, and some of its characteristics. A comprehensive discussion of it is found in the Appendix to this chapter. We begin with a definition. For positive prices and a feasible output vector ($u \in Rang P$) the *Cost Function* is

(2.2.3) $C(u, p) = \min_{x} \{px : x \in L(u)\}.$

The cost function expresses the minimum cost $px = (p_1 x_1 + \cdots + p_N x_N)$ at which a feasible output can be produced. (Since $p > 0$, and u is feasible, i.e., $L(u) \neq \emptyset$, the minimum in (2.2.3) is attained. See Appendix.) It is homogeneous of degree $+1$ in prices, since with $\lambda > 0$,

$$
\begin{aligned}
C(u, \lambda p) &= \min_{x} \{(\lambda p)x : x \in L(u)\} \\
&= \lambda \min_{x} \{px : x \in L(u)\} = \lambda C(u, p).
\end{aligned}
$$

At this point, homogeneity is the only property required. Next, consider the sets

(2.2.4) $\{u : C(u, p) \leqq C\} = \{u : C(u, p/C) \leqq 1\},$

where the equality follows from homogeneity of $C(u, p)$ in prices. The following proposition relates the two expressions (2.2.2) and (2.2.4).

(2.2.5)   **Proposition** :   Given that $p > 0$, then $\{u : u \in P(x), px \leqq C\} = \{u : C(u, p) \leqq C\}.$

Proof: Let $v \in \Re_+^M$, and $C(v, p) \leqq C$. Since $p > 0$, there exists an input vector x in $L(v)$ such that $C(v, p) = px$. Thus $v \in P(x)$, and since $px \leqq C, v \in \{u : u \in P(x), px \leqq C\}$. Conversely, if $u \in P(x)$ and $px \leqq C$ then $C(v, p) \leqq px \leqq C$, and hence $v \in \{u : C(u, p) \leqq C\}$.

Q.E.D.

Proposition (2.2.5) shows that for strictly positive prices, $p > 0$, the two expressions (2.2.2) and (2.2.4) may be used interchangeably to define the cost indirect output set. In later chapters on efficiency measurement and productivity gauging we thus have a choice in the representation of technology.

If prices $p \in \Re_+^N$ are not strictly positive, we then define the cost function as an infimum rather than a minimum, see the Appendix. Although it is still true that

14

$\{u : u \in P(x), px \leqq C\} \subseteq \{u : C(u,p) \leqq C\}$, a Cobb-Douglas production function $u = x_1^\alpha x_2^{1-\alpha}, 0 < \alpha < 1$, can be used to show that equality need not hold.

The cost indirect technology inherits properties from the parent technology. In deriving these properties, we assume that $C > 0$, since the case where C equals zero is not economically interesting. We begin with

> IP.1   (a) $0 \in IP(p/C), \forall (p/C) \in \Re_+^N$;   (b) $IP(0) = Rang\,P$.

To prove (a), let $(p/C) \in \Re_+^N$, and consider the definition $IP(p/C) = \{u : u \in P(x), px \leqq C\}$. Clearly, the zero input vector satisfies the budget (cost constraint), thus by Axiom P.1(a), $0 \in IP(p/C)$. To prove (b), let $p = 0$, then the budget set $B(p/C) = \{x : px \leqq C\} = \Re_+^N$, which equals $Dom\,P$. Now, $Rang\,P = \{u \in \Re_+^M : \exists x \in \Re_+^N \text{ such that } u \in P(x)\}$, and since $p = 0, px \leqq C$. Hence $u \in IP(0)$. Conversely if $u \in IP(0)$, then $u \in P(x)$ for some $x \in B(0) = \Re_+^N$ implying that $u \in Rang\,P$.

Q.E.D.

The (a) property states that inactivity is possible, and (b) shows that if inputs are free, any feasible output can be obtained, i.e., the budget constraint is not binding.

> IP.2 $IP((p/C)^\circ) \subseteq IP(p/C)$ if $(p/C)^\circ \geqq (p/C)$.

This axiom models monotonicity of the indirect output correspondence. It states that if one cost deflated price vector is at least as large as another, the output set associated with the former is included, not necessarily strictly, in the latter. Thus the indirect output correspondence is monotonically nonincreasing in input prices. The direct output correspondence on the other hand is nondecreasing in inputs quantities. To prove IP.2 we need merely observe that $B((p/C)^\circ) \subseteq B(p/C)$, i.e., the input choices at (p/C) include those at $(p/C)^\circ$.

> IP.3   If $u \in IP(p/C)$ and $0 \leqq \theta \leqq 1$, then $\theta u \in IP(p/C)$.

Weak disposability of outputs is modeled by IP.3. This axiom coincides with P.3, and tells us that proportional reductions of output can be achieved without violating the budget constraint. The proof is immediate from P.3 and the definition of $IP(p/C)$.

> IP.3.S   If P.3.S holds, $u \leqq v \in IP(p/C) \Longrightarrow u \in IP(p/C)$.

Strong disposability is modeled by IP.3.S. Again, this axiom is analogous to the parent axiom P.3.S, and assures that each output can be disposed of without violating the budget constraint, i.e., disposable is costless. The proof is left to the reader.

IP.4   If $(p/C) > 0, IP(p/C)$ is bounded.

A set is bounded if there exists a ball of finite radius that contains it. IP.4 tells us that if all prices are positive, and the target cost C is finite, then only finite amounts of outputs can be produced. IP.4 may be deduced from P.4, however we use property C.6 of the cost function (see Appendix) to prove IP.4. Since $p > 0$ and $0 < C < +\infty$ are given, if $IP(p/C)$ is not bounded then there is a sequence $u^\ell \in IP(p/C)$ such that $\| u^\ell \| \longrightarrow +\infty$ as $\ell \longrightarrow +\infty$. But by C.6, $C(u^\ell, p) \longrightarrow +\infty$ as $\| u^\ell \| \longrightarrow +\infty$, contradicting $C(u^\ell, p) \leqq C \, \forall \ell$.

Q.E.D.

If prices are not strictly positive we can show that $IP(p/C)$ need not be bounded. Consider a Cobb-Douglas production function of the simplest kind, $u = x_1^{\frac{1}{2}} x_2^{\frac{1}{2}}$. Choose $p_1 = 1 = C$ and $p_2 = 0$. The set $IP(1,0)$ is then $\{u \in \Re_+ : u \leqq (x_1 x_2)^{\frac{1}{2}}, x_1 \leqq 1, x_2 \geqq 0\}$. Take $x_1 = 1$, then as x_2 goes from 0 to $+\infty$, the above output set becomes $[0, +\infty)$ and is therefore unbounded.

IP.5   If $(p/C) > 0, IP(p/C)$ is closed.

Property IP.5 states that for each $(p/C) > 0$, the indirect output set is closed. Together with IP.4, the output set is thus also compact, and with IP.1(a) it is a nonempty compact set. Such sets are ideal for optimization, since a continuous function achieves a maximum and a minimum on that type of set. The proof of IP.5 follows directly from C.7 on the cost function (see the Appendix), since a function is lower semi-continuous if and only if its lower level sets are closed. Specifically, if $(p/C) > 0, IP(p/C) = \{u : C(u,p) \leqq C\}$ is closed since $C(u,p)$ is lower semi-continuous in u.

IP.6   If $u \in Rang P$, then $\forall p \in \Re_+^N$ there exists a scalar θ such that $u \in IP(\theta p/C)$.

This axiom states that if u is feasible, then by scaling target costs, or equivalently, proportionally scaling input prices, then that output can be produced with inputs that satisfy the budget constraint. The proof is omitted.

IP.7   If P.9 holds, then $IP(p/C)$ is convex $\forall (p/C) > 0$.

16

If the graph is convex then according to IP.7, the indirect output sets are also convex.

Proof: Suppose $u, v \in IP(p/C)$ then there exists x and y such that $u \in P(x), v \in P(y)$ with px and $py \leqq C$. By P.9, $\lambda u + (1 - \lambda)v \in P(\lambda x + (1 - \lambda)y)$ if $0 \leqq \lambda \leqq 1$, also, $\lambda px + (1 - \lambda)py \leqq C$, hence $IP(p/C)$ is convex.

$$Q.E.D.$$

The convexity of the indirect output set is a necessary condition for our later proof of duality, and here we will consider a condition weaker than P.9 that implies convexity of $IP(p/C)$.

P.10 $\forall x, y \in \Re^N_+$ and $\theta \in [0, 1]$, $\exists \lambda \in [0, 1]$ such that $\theta P(x) + (1 - \theta)P(y) \subseteqq$

$$P(\lambda x + (1 - \lambda)y).$$

Note that in contrast to P.9, two different scalars λ and θ are used here.

To prove that P.10 implies that $IP(p/C)$ is convex for all $(p/C) > 0$, assume that $u, v \in IP(p/C)$. Then there exist x, y such that $u \in P(x)$ and $v \in P(y)$ with $px \leqq C$ and $py \leqq C$. By P.10, $\theta u + (1 - \theta)v \in P(\lambda x + (1 - \lambda)y)$ for some $\lambda \in [0, 1]$, and $\lambda px + (1 - \lambda)py \leqq C$, therefore $\theta u + (1 - \theta)v \in IP(p/C)$.

$$Q.E.D.$$

2.3   The Parent and Indirect Technologies: A First Comparison

The above two sections show that certain similarities exist between the parent output set $P(x)$ and the indirect (cost constrained) output set $IP(p/C)$. The null vector belongs to both and output disposability properties are the same. Also if prices are strictly positive, the indirect output set is compact as is the parent or direct output set. Monotonicity differs, however. Whereas the direct output correspondence is nondecreasing in inputs, the indirect correspondence is nonincreasing in (p/C), reflecting the fact that as input prices increase, the affordable set becomes smaller.

Since both $P(x)$ and $IP(p/C)$ are output sets, we can compare them diagramatically. Suppose that an input vector $x \in \Re^N_+$ satisfies the budget constraint $px \leq C$ for some prices p and target cost C. Then $P(x) \subseteqq IP(p/C)$, with the indirect correspondence being the union of all direct output sets that use inputs at a cost less than or equal to the target cost C. Figure 2.4 illustrates.

17

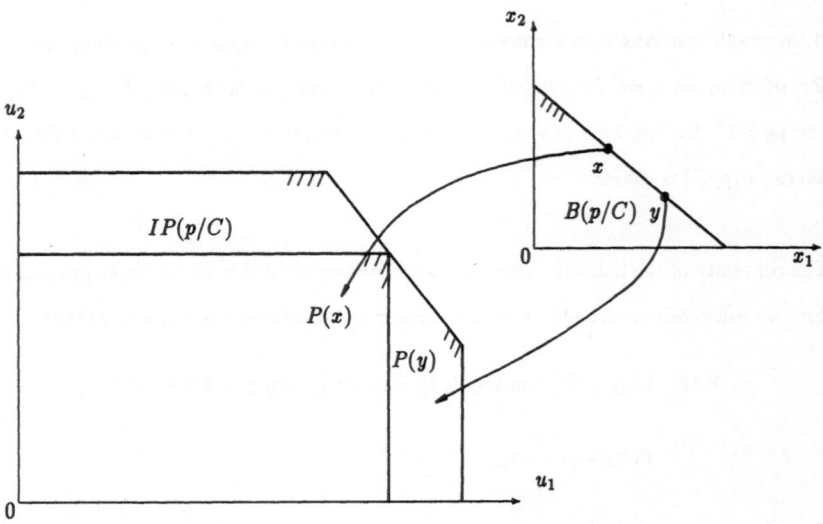

Figure 2.4: The Direct and Indirect Output Sets

In Figure 2.4 two different direct output sets and one indirect output set are drawn. The two sets $P(x)$ and $P(y)$ are direct input sets for which the total cost of inputs do not exceed C, i.e., $px \leq C$ and $py \leq C$. Thus these two sets are subsets of the indirect output set $IP(p/C)$. In fact, the indirect output set is the union of all direct output sets which can be produced within the budget constraint C. Formally $IP(p/C) = \cup_{px \leq C} P(x)$.

Next we would like to provide a preview of the dual relationship between the direct and indirect technologies. We begin with the scalar output case, which is familiar in economics both from production and utility theory. The *Production Function* representing the single output technology is defined by

$$(2.3.1) \qquad F(x) = \max\{u : u \in P(x)\}.$$

When u is a scalar, axioms P.1, P.3, P.4 and P.5 result in $P(x)$ being a compact interval containing the zero vector. The compactness suffices for a maximum to exist in (2.3.1), and for $F(x)$ to be appropriately defined as the largest element in that interval.

Corresponding to the direct production function above, we may define the *Cost Indirect*

18

Production Function as

(2.3.2) $IF(p/C) = \max\{u : u \in IP(p/C)\} = \max\{u : u \in P(x), px \leqq C\},$

where $IF(p/C)$ is the production side counterpart of the indirect utility function. Whenever $(p/C) > 0$, we have shown that $IP(p/C)$ is a compact set, thus the indirect production function is well defined. The duality between these two functions may be expressed as

$$(2.3.3) \qquad \begin{aligned} IF(p/C) &= \max_{x}\{F(x) : px \leqq C\} \\ F(x) &= \min_{(p/C)}\{IF(p/C) : px \leqq C\}. \end{aligned}$$

We do not prove this duality theorem at this point, although a few remarks are in order. The indirect production function is obtained from the search for the maximal scalar output that can be produced under the budget constraint $px \leqq C$. Correspondingly, the direct production function is obtained as the cheapest way, minimizing over cost deflated prices, to produce output. The duality pair (2.3.3) shows that each function can be retrieved using information on the other. The more general relationship stated in terms of distance functions (which generalize to the multiple output case) is proved in Chapter 5.

It is also of interest to establish a (preliminary) relationship between the indirect production function and the cost function. Assume that $(p/C) > 0$, so that the indirect output correspondence may be defined as

(2.3.4) $IP(p/C) = \{u : C(u, p/C) \leqq 1\}.$

Also assume that *Constant Returns to Scale (CRS)* prevails, i.e., $C(\theta u, p/C) = \theta C(u, p/C)$, where $\theta > 0$. If a single output is produced, i.e., $M = 1$, then the CRS cost function takes the simple form (see Problem 2.P.8)

(2.3.5) $C(u, p/C) = uC(1, p/C).$

Now, insert (2.3.5) into (2.3.4) and apply the definition of $IF(p/C)$, (2.3.2). This yields

$$(2.3.6) \qquad \begin{aligned} IF(p/C) &= \max\{u : uC(1, p/C) \leqq 1\} \\ &= \max\{u : u \leqq 1/C(1, p/C)\} \\ &= 1/C(1, p/C). \end{aligned}$$

Thus we have shown that if scalar output is produced and the technology exhibits constant returns to scale, then the indirect production function equals the reciprocal of the cost function.

2.4 The Revenue Restricted Technology

The revenue restricted technology or, as it is also termed, the revenue indirect technology models the set of input vectors which, given output prices $r = (r_1, \cdots, r_M) \in \Re_+^M$, can yield at least given positive target revenue $R > 0$. The *Revenue Indirect Input Correspondence* is defined by

$$(2.4.1) \qquad IL : \Re_+^M \longrightarrow IL(r/R) \subseteq \Re_+^N$$

where

$$(2.4.2) \qquad IL(r/R) = \{x : x \in L(u), \sum_{m=1}^{M} r_m u_m \geqq R\} = \{x : x \in L(u), (r/R)u \geqq 1\}$$

denotes the *Revenue Indirect Input Set*. This input set contains all input vectors that can produce revenue R given output prices r. As in the case of the indirect output correspondence, the indirect input correspondence may be defined in two ways. To formulate the second, we first must introduce the *Revenue Function*. The revenue function models the maximal revenue $ru = (r_1 u_1 + \cdots + r_M u_M)$ that can be attained with a given input vector $x \in \Re_+^N$ and fixed output prices $r \in \Re_+^M$, formally

$$(2.4.3) \qquad R(x, r) = \max_u \{ru : u \in P(x)\}.$$

The maintained axioms imply that $P(x)$ is a compact nonempty subset of $\Re_+^M$, and since ru is a continuous function, the maximum over all feasible output vectors is achieved. That is, for some $u^* \in P(x), R(x, r) = ru^*$, where u^* is dependent on (x, r). The second formulation of the indirect input correspondence, $R > 0$, is

$$(2.4.4) \qquad \{x : R(x, r) \geqq R\} = \{x : R(x, r/R) \geqq 1\}.$$

The relation between (2.4.2) and (2.4.4) is revealed in

$(2.4.5)$ **Proposition:** For $(r/R) \in \Re_+^M$, and $R > 0$,
$$\{x : x \in L(u), ru \geqq R\} = \{x : R(x, r) \geqq R\}.$$

Proof: For all $r \in \mathfrak{R}_+^M$, such that $ru < R(R > 0)$ for $u \in \mathfrak{R}_+^M$, the two sets are empty and equal. Thus assume that $y \in L(u)$ for some u such that $ru \geq R$, then $u \in P(y)$. Since the revenue function is the maximum over all $u \in P(y), R(y,r) \geq ru \geq R$. Thus $y \in \{x : R(x,r) \geq R\}$. Conversely, if $y \in \mathfrak{R}_+^N$ and $R(y,r) \geq R$, then $R(y,r) = ru^* \geq R$ for some $u^* \in P(y)$. Thus $y \in L(u^*)$ and $ru^* \geq R$, therefore $y \in \{x : x \in L(u^*) \text{ and } ru^* \geq R\}$.

$$Q.E.D.$$

Note that Proposition (2.4.5) does not require that output prices be strictly positive or even semi-positive. (Compare (2.2.5)).

Under the maintained axioms, the indirect input correspondence inherits related properties from the parent technology. We discuss these properties next, and assume throughout that $R > 0$.

IL.1   (a) $IL(0) = \emptyset$;   (b) $0 \notin IL(r/R), r \in \mathfrak{R}_+^M$.

Property (a) states that if output prices equal zero, then the indirect input set is empty. This follows from the fact that when $r = 0$ and $R > 0$, there is no $u \in \mathfrak{R}_+^M$ such that $ru \geq R$. The second part of IL.1 claims that $0 \notin IL(r/R)$ for any $r \in \mathfrak{R}_+^M$. If r is such that $ru < R$ for $u \in \mathfrak{R}_+^M$, then $IL(r/R)$ is empty and $0 \notin IL(r/R)$. If on the other hand $ru \geq R$, then $u \geq 0$, and by $IL(r/R) = \{x : x \in L(u), ru \geq R\} = \{x : u \in P(x), ru \geq R\}$, it follows from P.1(b) that $x \geq 0$. This proves that $0 \notin IL(r/R)$. Note that in the proof of (b), the property P.1(b) on the output set was used. However, IL is defined in terms of the input set $L(u)$. The resolution to this puzzle is of course that $L(u)$ and $P(x)$ are inverses, and therefore, $L(u)$ permits a set of axioms that are equivalent to those of $P(x)$.

IL.2 $x \in IL(r/R)$ and $\lambda \geq 1 \Longrightarrow \lambda x \in IL(r/R)$.

IL.2.S   If P.2.S ($\Longleftrightarrow$ L.2.S) holds, $y \geq x \in IL(r/R) \Longrightarrow y \in IL(r/R)$.

Disposability of inputs are modeled by the last two properties; weak disposability by IL.2 and strong disposability by IL.2.S. Weak disposability captures the idea that if x can produce target revenue at output prices r, then so can any scalar expansion of it. Strong disposability states that in addition to proportional expansions, any input vector at least as large as x can also yield revenue R. We prove IL.2 and leave IL.2.S to the reader. Assume that $x \in IL(r/R)$, i.e., $x \in L(u)$

21

for some u such that $ru \geq R$. Since by the maintained axioms, L satisfies weak disposability of inputs, $\lambda x \in L(u)$ if $\lambda \geq 1$. Thus $\lambda x \in IL(r/R)$.

IL.3 $IL(r/R) \subseteq IL((r/R)^\circ)$ if $(r/R)^\circ \geq (r/R)$.

Monotonicity is expressed by IL.3. Specifically, the indirect input correspondence is nondecreasing in revenue deflated output prices. That is, if output prices increase (holding target R constant), the associated set of outputs satisfying the revenue constraint requires fewer inputs. In contrast, the direct input correspondence is nonincreasing in input quantities. That is, the monotonicity is reversed between the direct and indirect formulations. Here we prove monotonicity of $L(u)$. Assume that outputs are strongly disposable, that is, $v \geq u \in P(x) \Longrightarrow v \in P(x)$, see P.3.S. Let $u \in P(x)$ and $u \geq v$, then $x \in L(u) \Longrightarrow x \in L(v)$, i.e., $L(u) \subseteq L(v)$. Conversely, assume there is an input vector $x \in \Re_+^N$ such that $u \in P(x)$ but $v \notin P(x)$, where $u \geq v$. Then $x \in L(u)$ but $x \notin L(v)$. This proves that P.3.S is equivalent to L.3.S which reads

L.3.S $\forall u, v \in \Re_+^M, u \geq v \Longrightarrow L(u) \subseteq L(v)$.

The following property expresses the idea of boundedness, that is if R becomes large, i.e., $R^\ell \longrightarrow +\infty$, then $\cap_{\ell=1}^{+\infty} IL(r/R^\ell)$ is empty. To understand this property, again look at P.4. Formally IL.4 reads

IL.4 If $R^\ell \longrightarrow +\infty$ as $\ell \longrightarrow +\infty$, $\displaystyle\bigcap_{\ell=1}^{+\infty} IL(r/R) = \emptyset$.

Next we discuss

IL.5 $IL(r/R)$ is closed.

If $IL(r/R)$ is empty then it is closed, therefore we need only consider output prices r such that $ru \geq R$. Let x^ℓ be a convergent sequence $x^\ell \longrightarrow x^\circ$, such that $x^\ell \in IL(r/R)$ for all ℓ. We need to prove that $x^\circ \in IL(r/R)$. Whenever $x^\ell \in IL(r/R)$ there exists a sequence u^ℓ such that $x^\ell \in L(u^\ell)$ and $ru^\ell \geq R$ for all ℓ. If u^ℓ does not converge, i.e., $\| u^\ell \| \longrightarrow +\infty$ as $\ell \longrightarrow +\infty$, then since x^ℓ is convergent and therefore bounded, P.4 fails. Thus u^ℓ is convergent. Now since $P(x)$ is a closed correspondence if and only if $L(u)$ is a closed correspondence, $x^\circ \in L(u^\circ)$, where u° is the limit point of u^ℓ. Moreover, since ru is continuous, $ru^\ell \longrightarrow ru^\circ \geq R$. Therefore, $x^\circ \in IL(r/R)$, and IL.5 holds.

IL.6 If $x \in L(u), u \geq 0$, then there exists a scalar λ such that $x \in IL(\lambda r/R)$.

Here attainability is modeled in the sense that if x can produce u, $u \geq 0$, then by scaling revenue or output prices, x can produce R.

IL.7 If P.9 holds, then $IL(r/R)$ is convex.

This property shows that a convex graph yields a convex indirect input set. A second condition on the parent technology that makes $IL(r/R)$ convex is

P.11 $\forall x, y \in \Re_+^N$ and $\forall \lambda \in [0,1]$, $\exists \theta \in [0,1]$ such that $\theta P(x) + (1-\theta)P(y) \subseteqq$

$$P(\lambda x + (1-\lambda)y).$$

Prior to proving our claim, it is important to note the differences between P.10 and P.11. In P.10, there exists a λ between zero and one that constructs an average of x and y, whereas in P.11 there exists a $\theta \in [0,1]$ that averages $P(x)$ and $P(y)$. Now for the proof of our claim. Let $u, v \in IL(r/R)$ and $\theta \in [0,1]$ (if $IL(r/R)$ is empty we need not prove anything). Then $u \in P(x)$ and $v \in P(y)$ for some $x, y \in \Re_+^N$ and $ru \geqq R$, $rv \geqq R$. By P.11 there exists $\lambda \in [0,1]$ such that $\theta u + (1-\theta)v \in P(\lambda x + (1-\lambda)y)$. Thus since $\theta ru + (1-\theta)rv \geqq R$, the claim is proved.

We note that although we have not derived all of the properties which the direct input correspondence L possesses due to its inverse relation to the direct output correspondence P, one may draw parallels between such properties on L and inherited properties on IL. We leave this task to the reader.

To illustrate the relationship between L and IL, note first that if u is feasible, i.e., $u \in RangP$, and $ru \geqq R$ then $L(u) \subseteqq IL(r/R)$. In Figure 2.5 we have drawn some direct input sets and one indirect input set under the condition that outputs meet the revenue constraint. The figure illustrates the idea that the indirect set envelops the set of direct input sets which satisfy the revenue constraint.

Appendix: Cost and Revenue Functions

Suppose input prices $p \in \Re_+^N$ are positive, i.e., $p > 0$, and that $u \in RangP$, i.e., u is feasible. Then the *Cost Function* is defined as

(2.A.1) $C(u,p) = \min_x \{px : x \in L(u)\} = \min_x \{px : u \in P(x)\}$,

23

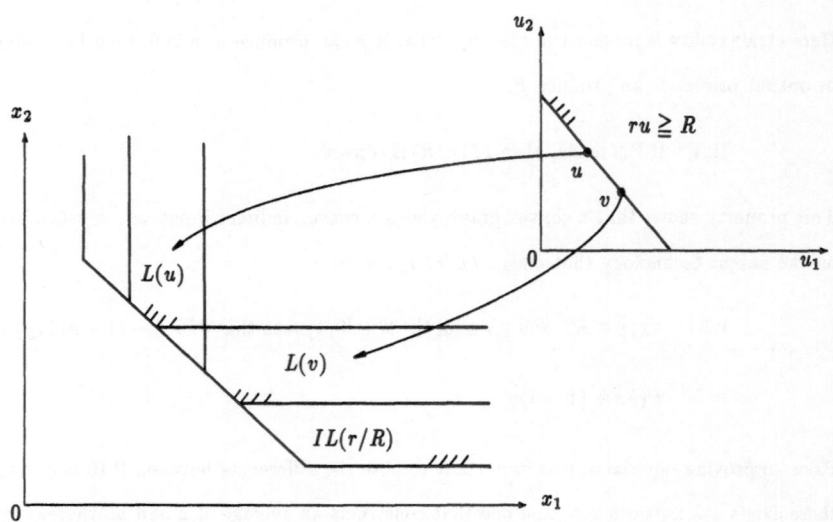

Figure 2.5: The Direct and Indirect Input Sets

where $px = \sum_{n=1}^{N} p_n x_n$. The equality holds between the two bracketed expressions since $x \in L(u) \iff u \in P(x)$, see (2.1.6). To show that a minimum is achieved, note that since the graph is closed, P.5, it follows that $L(u)$ is a closed set. Moreover $L(u)$ is nonempty, since $u \in Rang P$. By assumption $p > 0$; for $x^\circ \in L(u)$ the set $L(u) \cap \{x : px \leqq px^\circ\}$ is nonempty and compact, which is sufficient when a minimum (or a maximum) is sought. In our case we note that if $x' \notin L(u) \cap \{x : px \leqq px^\circ\}$, then $px' > px^\circ$, and a minimum is achieved on $L(u)$.

The properties on the cost function implied by the maintained axioms on the technology are:

C.1 $C(u, p)$ is nonnegative and nondecreasing in (positive) prices.

C.2 $C(u, \lambda p) = \lambda C(u, p), \lambda > 0, p > 0$.

C.3 $C(u, p)$ is concave and continuous in (positive) prices.

C.4   (a) $C(0, p) = 0, \forall p > 0$; (b) $C(u, p) > 0, p > 0, u > 0$.

C.5 $C(\theta u, p) \leqq C(u, p), 0 \leqq \theta \leqq 1, p > 0$.

C.6   If $\| u^\ell \| \longrightarrow +\infty$ as $\ell \longrightarrow +\infty$, and $p > 0, C(u^\ell, p) \longrightarrow +\infty$ as $\ell \longrightarrow +\infty$.

C.7 $C(u, p)$ is lower semi-continuous in u, for $p > 0$, see A.6 in Appendix A.

24

Proof: The proof of C.1 is left to the reader. C.2 is proved in Section 2.2, thus consider C.3. Suppose p° and p' are positive and that $0 \leqq \lambda \leqq 1$. Then

$$C(u, p^\circ) = p^\circ x^\circ, \text{ where } x^\circ \in L(u) \text{ and}$$
$$C(u, p') = p'x' \text{ where } x' \in L(u).$$

Moreover,

$$
\begin{aligned}
C(u, \lambda p^\circ + (1 - \lambda)p') &= (\lambda p^\circ + (1 - \lambda)p')x^*(\lambda), \text{ where } x^*(\lambda) \in L(u) \\
&\geqq \lambda p^\circ x^\circ + (1 - \lambda)p'x' \\
&= \lambda C(u, p^\circ) + (1 - \lambda)C(u, p').
\end{aligned}
$$

This proves that the cost function is concave in input prices. The proof that a concave function is continuous on an open set, i.e., on $\{p : p > 0\}$, can be found in Rockafeller (1970, p. 82).

C.4   follows from P.1 (a) and (b) respectively.

To prove C.5, note first that P.3 is equivalent to

L.3 $L(u) \subsetneqq L(\theta u), 0 \leqq \theta \leqq 1.$

By L.3 we have $\min_x\{px : x \in L(\theta u)\} \leqq \min_x\{px : x \in L(u)\}.$

Property C.6 states that if output becomes large cost becomes large. To prove that C.6 applies, we note that by P.4, $P(x)$ is bounded for all $x \in \Re_+^N$. Thus if $\| u^\ell \| \longrightarrow +\infty$ as $\ell \longrightarrow +\infty$, there must be a sequence x^ℓ and $\| x^\ell \| \longrightarrow +\infty$ in order for u^ℓ to be feasible, thus since $p > 0, C(u^\ell, p) \longrightarrow +\infty$ as $\ell \longrightarrow +\infty$.

To prove that the cost function is lower semi-continuous in output whenever $p > 0$, C.7, let $u^\ell \longrightarrow u^\circ$ be an arbitrarily chosen sequence of feasible outputs. Consider the sequence $C(u^\ell, p) = px^*(u^\ell, p)$, where $x^*(u^\ell, p)$ is the cost minimizing input bundle. Each term of $x^*(u^\ell, p)$ is bounded since $p > 0$. If $\{px^*(u^\ell, p) : \ell = 1, 2, \cdots\}$ is not a bounded set, then $\liminf_{\ell \longrightarrow +\infty} C(u^\ell, p) \geqq C(u^\circ, p)$ and the cost function is lower semi-continuous in u. Therefore suppose that $\{px^*(u^\ell, p) : \ell = 1, 2, \cdots\}$ is a bounded set. Since $p > 0$, there is a subsequence $x^*(u^{\ell_k}, p) \longrightarrow x^\circ$. Since $x^*(u^{\ell_k}, p) \in L(u^{\ell_k})$ for all ℓ_k, and $L(u)$ is closed (P.5), $x^\circ \in L(u^\circ)$. Therefore, $px^\circ(u^\circ, p) \geqq C(u^\circ, p)$ and $\liminf_{\ell \longrightarrow +\infty} C(u^\ell, p) \geqq C(u^\circ, p)$.

Q.E.D.

25

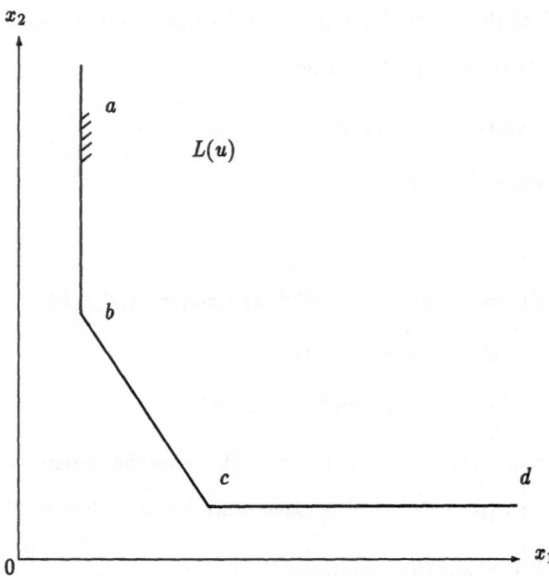

Figure 2.6: The Input Set and Its Efficient Subset

Next we extend the definition of the cost function to allow for nonnegative prices. Our first extension is

(2.A.2) $C(u,p) = \inf_{x}\{px : x \in L(u)\}, p \geq 0, u \in RangP.$

Clearly, if $p > 0$, (2.A.2) and (2.A.1) coincide, also whenever there exists $x \in L(u)$ such that $C(u,p) = px$, then the two are equal. However if there is no $x \in L(u)$ such that $C(u,p) = px$, then they differ. Take the Cobb-Douglas example $u = x_1^{\frac{1}{2}}x_2^{\frac{1}{2}}$ and assume that $p_1 = 0$, but that $p_2, u > 0$. Definition (2.A.1) no longer holds, but (2.A.2) yields $C(u,p) = 0$, although of course there is no $(x_1, x_2) \in L(u) = \{(x_1, x_2) : u \leq x_1^{\frac{1}{2}}x_2^{\frac{1}{2}}\}$, yielding this value. However, $C(u,p)$ may be approached as the limit of a sequence.

A second way to modify Definition (2.A.1) to allow for nonnegative prices is to impose restrictions on the input sets $L(u), u \in \Re_+^M$. Define the *Efficient Input Set* as

(2.A.3) $EffL(u) = \{x \in L(u) : y \leq x \Longrightarrow y \notin L(u)\}, u \geq 0$ and $EffL(0) = \{0\}.$

In Figure 2.6, the input set $L(u)$ is bounded by the line segments $(abcd)$. The corresponding efficient input set consists of $[bc]$. One may prove that if $L(u)$ is nonempty, so too is $EffL(u)$.

26

The proof hinges on the closedness of $L(u)$.

The restriction imposed on the technology which allows for nonnegative prices in defining the cost function as a minimum may now be introduced

(2.A.4) $EffL(u)$ is bounded for all $u \in \Re_+^M$.

An example of a technology with bounded efficient input sets is the Leontief technology given by

(2.A.5) $L(u) = \{(x_1, x_2) : u \leqq \min\{x_1, x_2\}\}$.

The Cobb-Douglas technology $L(u) = \{(x_1, x_2) : u \leqq x_1^{\frac{1}{2}} x_2^{\frac{1}{2}}\}$, on the other hand, has unbounded efficient input sets.

One more result is needed to guarantee that a minimum cost exists when prices may be zero, namely

(2.A.6) $L(u) \subsetneqq \overline{EffL(u)} + \Re_+^N$.

The proof of (2.A.6) is left to the reader. Now, consider the cost function (2.A.2) with bounded efficient subsets, then for $p \geq 0$, and $u \in RangP$,

$$
\begin{aligned}
(2.A.7) \quad C(u, p) &= \inf_x \{px : x \in L(u)\} \\
&\geqq \inf_x \{px : x \in \overline{EffL(u)} + \Re_+^N\} \\
&= \min_x \{px : x \in \overline{EffL(u)}\}.
\end{aligned}
$$

$\overline{EffL(u)}$ is nonempty and compact, thus px attains its minimum on $\overline{EffL(u)}$. Also, since $L(u)$ is closed, $\overline{EffL(u)} \subsetneqq L(u)$, therefore $C(u, p) = \min_x \{px : x \in L(u)\}$.

It is frequently assumed that the cost function is differentiable in (positive) prices. A condition on the technology, sufficient for differentiability is given in

(2.A.8)   **Proposition:** Suppose that the technology satisfies (P.1 - P.6) and P.8. Let $u \geq 0$ and $u \in RangL$. If $x, y \in EffL(u), x \neq y \implies (\lambda x + (1 - \lambda)y) \notin EffL(u), 0 < \lambda < 1$, then $C(u, p)$ is differentiable in $p, p > 0$.

The condition imposed on the technology by Proposition (2.A.8) states that the efficient input set $EffL(u)$ consists of extreme points only. Examples of production technologies that satisfy this condition include the Leontief technology and technologies with strictly quasi-concave production functions, i.e., if $x, y \in \Re_+^N, x \neq y$, and $0 < \lambda < 1$ then

(2.A.9) $F(\lambda x + (1 - \lambda)y) > min\{F(x), F(y)\}.$

To prove Proposition (2.A.8) a lemma and a remark are needed. The remark is from Rockafellar.

(2.A.10)   **Remark:** By Rockafellar (1970, p. 242) it follows that the cost function is
             differentiable in $p, p > 0$ at (u, p) if and only if $CM(u, p)$ is a singleton.

The set

(2.A.11) $CM(u, p) = \{x \in L(u) : px = C(u, p)\}, u \in RangP, p \in \Re_+^N,$

is called the *Cost Minimization Set*, and it denotes all input vectors that minimize the cost of producing output u at prices p. The required lemma states

(2.A.12)   **Lemma :** $\displaystyle\bigcup_{p>0} CM(u, p) \subseteqq EffL(u), u \geq 0, u \in RangP.$

Proof: Let $u \geq 0, u \in RangP$ and $x \in CM(u, p)$ for some $p > 0$. Assume that $x \notin EffL(u)$, then there exists $y \in EffL(u)$ such that $y \leq x$. Thus, since $p > 0, py < px$. This contradicts the supposition $x \in CM(u, p)$.

Q.E.D.

Proof of Proposition (2.A.8). First if $EffL(u)$ is a singleton, then $CM(u, p)$ is a singleton for all $p > 0$, thus by the Remark, $C(u, p)$ is differentiable in $p, p > 0$. Suppose next that $x, y \in CM(u, p), x \neq y$ for some $p > 0$, i.e., $C(u, p)$ is not differentiable. By Lemma (2.A.12), $x, y \in EffL(u)$, moreover, $(p\lambda x + p(1 - \lambda)y) = px = py, 0 < \lambda < 1$ and by convexity of $L(u), (\lambda x + (1 - \lambda)y) \in L(u)$, but by hypothesis $(\lambda x + (1 - \lambda)y) \notin EffL(u)$. Thus there exists $x^\circ(\lambda) \in EffL(u)$ such that $x^\circ(\lambda) \leq (\lambda x + (1 - \lambda)y)$, therefore $px^\circ(\lambda) < px = py$. This contradicts $x, y \in CM(u, p)$.

Q.E.D.

We now turn our attention to the revenue function. Suppose that an output price vector $r \in \Re_+^M$ and an input vector $x \in \Re_+^N$ are given. The *Revenue Function* is defined as

(2.A.13) $R(x, r) = \max_u\{ru : u \in P(x)\}.$

The revenue function is the maximum revenue that can be obtained from an input vector x given that output prices are $(r_1, \cdots, r_M)$. It is defined as a maximum, which is appropriate since ru is a continuous function and the output set $P(x)$ is nonempty, $0 \in P(x)$ for all $x \in \Re_+^N$, and compact, P.4 and P.5. The revenue function satisfies properties similar to those of the cost function, including:

R.1 $R(x,r) \geq 0$ and nondecreasing in (positive) prices.

R.2 $R(x, \theta r) = \theta R(x,r), \theta > 0$.

R.3 $R(x,r)$ is convex and continuous in (positive) prices.

R.4 $R(0,r) = 0, \forall r \in \Re_+^M$.

R.5 $R(\lambda x, r) \geq R(x,r), \lambda \geq 1, r \in \Re_+^M$.

R.6   If $R(x,r) > 0$, then $R(\lambda x, r) \longrightarrow +\infty$ as $\lambda \longrightarrow +\infty$.

R.7 $R(x,r)$ is upper semi-continuous in x for all $r \in \Re_+^M$, see A.7 in Appendix A.

The proofs of properties (R.1 - R.6) are similar to the proofs of (C.1 - C.6) of the cost function. Thus we only prove R.7. Let x^ℓ be a convergent sequence, i.e., $x^\ell \longrightarrow x^\circ$, and consider $R(x^\ell, r)$. Since $x^\ell \longrightarrow x^\circ, R(x^\ell, r)$ is a bounded sequence, and there exists a subsequence $x^{\ell_k} \longrightarrow x^\circ$ with $ru(x^{\ell_k}, r) = R(x^{\ell_k}, r)$ and $u(x^{\ell_k}, r) \in P(x^{\ell_k})$. P is a closed correspondence and $R(x^{\ell_k}, r)$ is convergent, thus $u(x^\circ, r) \in P(x^\circ)$ and $\limsup_{k \to +\infty} R(x^{\ell_k}, r) = ru^\circ \leq R(x^\circ, r)$. This proves R.7.

In our discussion of the cost function, we provided a condition on the technology that is sufficient for differentiability of $C(u,p)$ in (positive) prices. A similar condition can be developed for the revenue function. Doing so, define

(2.A.14) $RM(x,r) = \{u \in P(x) : ru = R(x,r)\}$.

The set $RM(x,r)$ denotes all feasible output vectors which maximize revenue for (x,r). This set is called the *Revenue Maximization Set*. Now it may be proved that

(2.A.15) $\bigcup_{r>0} RM(x,r) \subseteqq EffP(x)$,

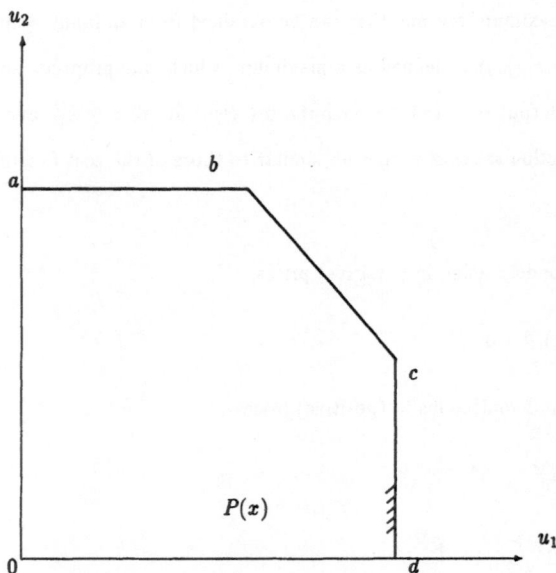

Figure 2.7: An Efficient Output Set

where $EffP(x)$ denotes the *Efficient Output Set* for x. This set is defined by

(2.A.16) $EffP(x) = \{u \in P(x) : v \le u \Longrightarrow v \notin P(x)\}, P(x) \ne \{0\}$, and $EffP(x) = 0$

if $P(x) = 0$.

Figure 2.7 illustrates. The output set $P(x)$ is bounded by $(0abcd0)$. The efficient output vectors are the line segment (bc), including points b and c. Note that the line segment (ab), for example, does not belong to $EffP(x)$.

From the above discussion it should be possible for the reader to formulate the revenue version of differentiability, i.e., the revenue version of (2.A.7).

2.P Problems

(2.P.1)   Suppose the technology transforms a single input $x \ge 0$ into a single output $u \ge 0$ according to $u = x^2$. Derive $P(2)$, $L(4)$ and the graph.

(2.P.2)   Which axioms does (a) $P(x) = \{u : u \le (x+2)^{\frac{1}{2}}\}$; (b) $P(x,y) = \{u : u \le (xy)^{\frac{1}{2}}\}$; (c) $P(x,y) = \{u : u \le (xy)^{\frac{1}{2}}/(1+x)\}$ satisfy?

(2.P.3)   Prove that (a) $DomL = RangP$, and that (b) $DomP = RangL$.

(2.P.4)   Suppose that the efficient input sets (see 2.A.3) are bounded for all $u \in RangP$. Prove
           that for $p \geq 0$, $\{u : C(u, p) \leq C\} \subseteq \{u : u \in P(x), px \leq C\}$.

(2.P.5)   Suppose inputs are strongly disposable. Prove IP.4 using definition (2.2.2).

(2.P.6)   Prove property IP.6.

(2.P.7)   Let $F(x)$ be a production function. Show that the output set $P(x) = \{u : u \leq F(x)\}$.

(2.P.8)   If the technology exhibits constant returns to scale, and a single output is produced,
           then the cost function may be written as $C(u, p) = u \cdot C(1, p)$. Verify this claim.

(2.P.9)   Prove IL.2.S.

(2.P.10)   Suppose that $L(u) \neq \emptyset$ and closed. Show that $EffL(u) \neq \emptyset$.

(2.P.11)   Give an example of a production technology (production function) for which
           $\bigcup_{p>0} CM(u, p)$ is a proper subset of $EffL(u)$.

(2.P.12)   Prove that $L(u) \subseteq \overline{EffL(u)} + \Re_+^N$.

(2.P.13)   Suppose that the production function equals $u = x_1 + x_2$. Find the corresponding cost
           function, and show that it is not differentiable.

(2.P.14)   Verify properties (R.1–R.6).

(2.P.15)   Show that $\bigcup_{r>0} RM(x, r) \subseteq EffP(x)$.

Notes

The axioms introduced on the parent technology are adopted from Shephard (1970) and
Shephard and Färe (1980). Additional axiomatic settings can be found in Debreu (1959), Fuss
and McFadden (1978), Koopmans (1951) and Teusch (1983). Eichhorn and Leopold (1990)
discuss logical aspects of Shephard's axioms, i.e., consistency and independence of the axioms.

The cost restricted scalar valued production (utility) function has been extensively studied in
economics, see Blackorby, Primont and Russell (1978). For applications see Chambers and Lee

31

(1986), Kim (1988) and Garofalo and Malhotra (1990). Here we follow Shephard (1974), Shephard and Färe (1980), and Färe (1988).

Chapter 3

Representations of Technology

3.0 Introduction

The purpose of this chapter is to introduce alternative representations of technology, with an emphasis on representations which are functions rather than representations which are sets or correspondences. We give particular attention to the distance function as a function representation of technology and show that it is a generalization (to the multiple output case) of the more familiar single output production function.

Section 3.1 introduces the direct and indirect output distance functions. The direct output distance function is a function representation of technology, specifically, it tells us "how far" any output vector in some $P(x)$ is from the boundary of that set; in the single output case the distance function can be interpreted as the ratio of actual to maximum potential output. This characteristic will prove extremely useful in gauging performance of individual firms or observations, which will be pursued in Chapter 6. As one might expect, the direct output distance function inherits properties from the parent technology, which is pursued in this section.

Since the focus of this manuscript is on restricted technologies, one may ask whether the cost indirect output correspondence has a function representation analogous to the role played by the direct output distance function with respect to the output correspondence. In response to this question, we introduce the indirect output distance function which is just such a function representation of the cost restricted technology. Like the direct output distance function, it expands a given output vector proportionately as much as is feasible while remaining in the reference set, which is in this case, the cost indirect output set. In the single output case this would correspond to the ratio of observed output to the maximum potential output achievable given a fixed budget, an interpretation which will be exploited to measure efficiency in Chapter 6. We explore the relationship of the indirect output distance function to the cost function as well as to the indirect production function and consider the properties it inherits from its parent technology.

Since one of our goals is to provide tools for those engaged in applied research, we consider two alternative parameterizations of these output distance functions, including a translog and a variation on the generalized Leontief functional form. These parameterized distance functions could be estimated in several ways: using ordinary least squares (both specifications can be made linear in the parameters), using a frontier composed error approach, or by employing a parametric linear programming approach similar to that used by Aigner and Chu (1968). Although deferred to Chapter 6, the direct and indirect distance functions could be calculated as solutions to nonparametric, linear programming problems, i.e., as reciprocals to Farrell type efficiency measures.

In Section 3.2 we turn to the function representations of the input set and the revenue indirect input sets (and their correspondences). Here we follow the same general approach as was taken in Section 3.1 in discussing function representations of the restricted and unrestricted output sets. We begin with the direct input distance function and show that it completely characterizes technology (specifically, an input vector is an element of $L(u)$ if and only if the input distance function takes on a value greater than or equal to unity). The input distance function takes the maximum radial contraction of an input vector while remaining in the input requirement set, i.e., it measures the distance to the boundary of the input requirement set, providing a natural measure of performance which will be exploited in Chapter 6. We explore the relationship between the input and output distance functions, in particular, we show that they are reciprocal to each other only when technology is linear homogeneous. We also discuss the properties inherited by the input distance function from the parent technology.

Section 3.2 continues with the introduction of the indirect input distance function which completely characterizes the revenue restricted technology, from which it inherits properties. This function representation of the revenue indirect technology measures the maximum radial contraction of inputs which is feasible given target revenue. This section concludes with translog and generalized Leontief type specifications of these input distance functions, emphasizing their

34

potential usefulness in empirical work.

The final section of this chapter is devoted to a different type of representation of technology. Here the goal is to provide piecewise linear representations of technology, where technology is a set. These piecewise linear representations of technology will prove useful in a linear programming framework, which we shall exploit in later chapters. In this section, however, we focus on specifying output and input sets as well as cost and revenue indirect technologies as a set of linear equations (which will serve as constraints in linear programming problems). We show that these satisfy the properties imposed by our set of maintained axioms, and show how to impose various restrictions to allow for weak and strong disposability as well as specifying returns to scale characteristics.

A set of problems and notes on the literature conclude the chapter.

3.1 The Direct and Indirect Output Distance Function

The intention of this section is to introduce function representations of the direct and indirect production correspondences. In the case of many inputs and outputs, the natural approach is to apply distance functions. In the direct case, these functions map input and output vectors into the real line, and in the indirect case they map cost deflated input prices and output quantities into a scalar. As we will demonstrate, in the case of a single output technology the direct and indirect output distance functions are closely related to the direct and indirect production functions, respectively. Also, since the indirect output distance function may be deduced from the cost function, and the cost function is dual to the direct input distance function which is "inverse" to the direct output distance function, see 3.2, we would also expect the two output distance functions to be dual. This connection is explored in Chapter 5.

In Chapter 2, the production function was defined as the maximal output that can be produced from an exogenously given input vector. The direct output distance function generalizes this notion to the multi-output case.

(3.1.1)   **Definition**: The function $D_o : \Re_+^N \times \Re_+^M \longrightarrow \overline{\Re}_+$ defined by $D_o(x, u) =$
$\inf\{\theta > 0 : (u/\theta) \in P(x)\}$ is called the *(Direct) Output Distance Function*.

In order to illustrate Definition (3.1.1), suppose that two outputs are produced (M = 2) and that

35

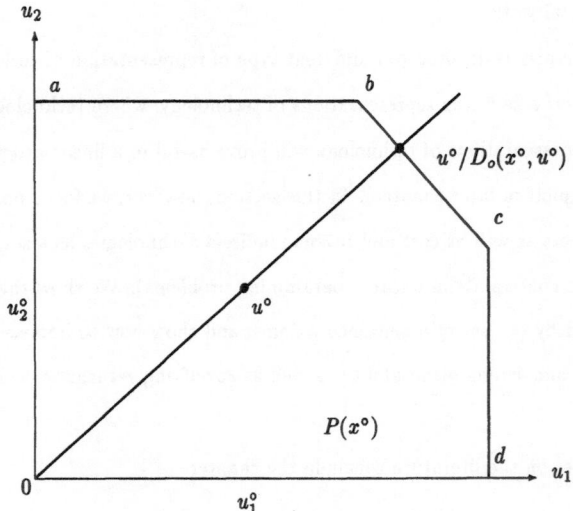

Figure 3.1: The Output Distance Function

(u_1^o, u_2^o) belongs to $P(x^o)$. This case is illustrated in Figure 3.1. The output set $P(x^o)$ is bounded by the line segments $(0abcd0)$, $u^o = (u_1^o, u_2^o)$ is feasible, $u^o \in P(x^o)$.

The distance function measures the maximal ray-expansion of u^o in $P(x^o)$. That is, in terms of distances,

$$(3.1.2) \qquad D_o(x^o, u^o) = \frac{\parallel u^o \parallel}{\parallel u^o/D_o(x^o, u^o) \parallel}$$

In order to make the connection between $D_o(x, u)$ and the production function $F(x)$, suppose N $= M = 1$, i.e., a single input is used to produce a single output. Figure 3.2 illustrates. In the example, $(x^o, u^o) \in GR$, so that $u^o \leqq F(x^o)$. The figure shows that $F(x^o) = u^o/D_o(x^o, u^o)$ or equivalently,

$$D_o(x^o, u^o) = u^o/F(x^o).$$

Thus in the scalar case the output distance function is merely the quotient between observed output u^o and maximal output $F(x^o)$. This result can be shown to hold even when $x^o \in \Re_+^N, N \geqq 1$.

To make sure that (3.1.1) is a valid definition of the output distance function we distinguish among four cases:

36

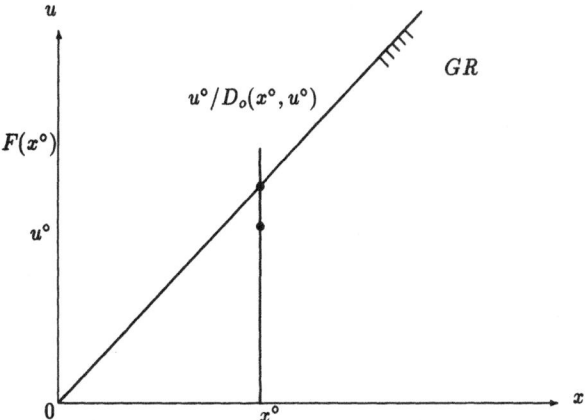

Figure 3.2: The Production and Output Distance Functions

P.a $u \geq 0$ and $u \in P(x)$,

P.b $u \geq 0, u \notin P(x)$, but there exists a $\theta > 0$ such that $\theta u \in P(x)$,

P.c $u \geq 0$ and there does not exist any $\theta > 0$ such that $\theta u \in P(x)$,

P.d $u = 0$.

By the maintained axioms, $P(x)$ is a compact set (P.4 and P.5). In case P.a, $u \in P(x)$, thus the infimum in Definition (3.1.1) is achieved and $D_o(x, u) < +\infty$. Moreover, since $u \in P(x), D_o(x, u) \leqq 1$ and since $u \geq 0, 0 < D_o(x, u)$. Thus in this case $0 < D_o(x, u) \leqq 1$. Case P.b models the situation where $u \notin P(x)$ but by taking a radial contraction of u, $\theta u \in P(x)$. In this case the infimum is also achieved but since $u \notin P(x)$, $1 < D_o(x, u) < +\infty$. In the third case, since $0 \in P(x), \forall x \in \Re_+^N$, but there is no $\theta > 0$ such that $\theta u \in P(x)$, $D_o(x, u) = +\infty$. In the final case, since infimum is used to define the output distance function, $D_o(x, 0) = 0$.

It is now clear that (3.1.1) appropriately defines the output distance function, however we need to show that the output correspondence can also be retrieved from the distance function. Thus define the induced output set

(3.1.3) $P_{D_o}(x) = \{u : D_o(x, u) \leqq 1\}, x \in \Re_+^N$.

37

The equivalence between $P(x)$ and $P_{D_o}(x)$ is proved in

(3.1.4)   **Proposition** : $P(x) = P_{D_o}(x), x \in \Re_+^N$.

Proof: Suppose $u \in P(x)$, then by the definition of $D_o(x, u)$, $u \in P_{D_o}(x)$. Conversely, assume that $u \notin P(x)$, then $D_o(x, u) > 1$ thus $u \notin P_{D_o}(x)$.

$$Q.E.D.$$

Proposition (3.1.4) proves that the parent and induced output sets are equal. Thus as a consequence we have also proved that the distance function completely characterizes technology.

(3.1.5)   **Corollary** : $u \in P(x)$ if and only if $D_o(x, u) \leqq 1, x \in \Re_+^N$.

In Chapter 2 a thorough discussion of the axioms on the parent technology was undertaken. Here we merely state and prove the properties that the direct output distance function inherits from $P(x)$.

$D_o.1$ (a) $D_o(x, 0) = 0, \forall x \in \Re_+^N$; (b) $D_o(0, u) = +\infty, \forall u \geq 0$.

$D_o.2$ $\forall (x, u) \in \Re_+^{N+M}, D_o(\lambda x, u) \leqq D_o(x, u), \lambda \geqq 1$.

$D_o.2.S$ $\forall u \in \Re_+^M$ and $x \geqq y, D_o(x, y) \leqq D_o(y, u)$.

$D_o.3$ $\forall (x, u) \in \Re_+^{N+M}, D_o(x, \theta u) = \theta D_o(x, u), \theta > 0$.

$D_o.4$ $\forall (x, u) \in \Re_+^{N+M}, D_o(x, \theta u) \leqq D_o(x, u), 0 \leqq \theta \leqq 1$.

$D_o.4.S$ $\forall x \in \Re_+^N$ and $u \geqq v, D_o(x, u) \geqq D_o(x, v)$.

$D_o.5$ (a) If $(x^\ell, u^\ell) \longrightarrow (x^o, u^o)$ and $D_o(x^\ell, u^\ell) \leqq 1$ for all ℓ, then $D_o(x^o, u^o) \leqq 1$; (b) $D_o(x, u)$ is lower semi-continuous on $\Re_+^M$, see A.6 in Appendix A; (c) $D_o(x, u)$ is lower semi-bounded on $\Re_+^M$, see A.14 in Appendix A.

$D_o.6$ If $D_o(x, u) \leqq 1$ and $u \geq 0$, then for all $\theta \geqq 0$ there exists a scalar λ_θ such that $D_o(\lambda_\theta x, u) \leqq 1/\theta$.

$D_o.7$ $\forall x \in \Re_+^N, D_o(x, u)$ is convex on $\Re_+^M$.

$D_o.8$ $\forall u \in \Re_+^M, D_o(x, u)$ is quasi $-$ convex on $\Re_+^N$, see A.8 in Appendix A.

The next proposition states that the properties (P.1 – P.8) on the output correspondence are equivalent to $(D_o.1 - D_o.8)$ on the output distance function.

38

(3.1.6)   **Proposition**: The output correspondence satisfies (P.1 – P.8) if and only if the
output distance function satisfies (D_o.1 – D_o.8).

Proof: (P.1(a) $\Longleftrightarrow$ D_o.1(a)) $0 \in P(x), \forall x \in \Re_+^N$ implies that
$D_o(x, 0) = \inf\{\theta > 0 : (0/\theta) \in P(x)\} = 0$. Conversely, $D_o(x, u) = 0 \Longrightarrow (u/\theta) \in P(x), \forall \theta > 0$, i.e.,
$u \in \theta P(x), \forall \theta > 0$ or $u \in \bigcap_{\theta > 0} P(x) = 0$, i.e., $u = 0$.

(P.1(b) $\Longleftrightarrow$ D_o.1(b)) Let $u \geq 0$ and suppose $u \notin P(0)$ then by the definition of
$D_o(x, u), D_o(0, u) = +\infty$. Conversely, if $D_o(x, u) > 1$ then $u \notin P(x)$. Thus if $D_o(0, u) = +\infty$ for
$u \geq 0$, it follows that $u \notin P(0)$.

(P.2 $\Longleftrightarrow$ D_o.2) Let $(x, u) \in \Re_+^{N+M}$ and $\lambda \geq 1$. It follows from P.2 that
$\inf\{\theta > 0 : (u/\theta) \in P(\lambda x)\} \leq \inf\{\theta > 0 : (u/\theta) \in P(x)\}$, i.e., $D_o(\lambda x, u) \leq D_o(x, u)$. The converse
is proved by $P(x) = \{u : D_o(x, u) \leq 1\} \subseteq \{u : D_o(\lambda x, u) \leq 1\} = P(\lambda x), \lambda \geq 1$.

(P.2.S $\Longleftrightarrow$ D_o.2.S) The proof is left to the reader.

(D_o.3) The proof of this property follows directly from the Definition (3.1.1) and is omitted.

(P.3 $\Longleftrightarrow$ D_o.3) and (P.3.S $\Longleftrightarrow$ D_o.4.S) are omitted.

(P.5 $\Longleftrightarrow$ D_o.5(a)) This equivalence follows from Corollary (3.1.5).

(P.5 $\Longrightarrow$ D_o.5(b)) Note that a function is lower semi-continuous only if its level sets are closed for
all values in the range. Let $\alpha = 0$, then $\{u : D_o(x, u) \leq 0\} = \{0\}$, thus let $\alpha > 0$. In this case,
$\{u : D_o(x, u) \leq \alpha\} = \alpha\{v : D_o(x, v) \leq 1\} = \alpha P(x)$. Thus since $P(x)$ is closed the implication is
proved.

(P.4 $\Longleftrightarrow$ D_o.5(c)) This property follows from $\{u : D_o(x, u) \leq \alpha\} = \alpha P(x)$.

(P.6 $\Longleftrightarrow$ D_o.6) Let $u \geq 0, x \geq 0$, then for all $\theta \geq 0$ there exists a scalar $\lambda_\theta \geq 0$ such that
$\theta u \in P(\lambda_\theta x)$, by P.6. Thus $D_o(\lambda_\theta x, \theta u) = \theta D_o(\lambda_\theta x, u)$. $\theta u \in P(\lambda_\theta x)$ implies that $D_o(\lambda_\theta x, \theta u) \leq 1$,
therefore $D_o(\lambda_\theta x, u) \leq 1/\theta$. The converse follows from the equality $P(x) = \{u : D_o(x, u) \leq 1\}$.

(P.7 $\Longrightarrow$ D_o.7) We prove this only for $u, v \in \Re_+^M, x \in \Re_+^N$ such that $D_o(x, u)$ and $D_o(x, v)$ are
positive and finite. By homogeneity, i.e., D_o.3, $D_o(x, v/D_o(x, v)) = 1$ and $D_o(x, u/D_o(x, u)) = 1$,

39

thus $v/D_o(x,v) \in P(x)$ as does $u/D_o(x,u)$. $P(x)$ is assumed to be convex thus $\lambda u/D_o(x,u) + (1-\lambda)v/D_o(x,v)$ also belongs to $P(x)$. Take $\lambda = D_o(x,u)/(D_o(x,u) + D_o(x,v))$, then $(u+v)/(D_o(x,u) + D_o(x,v)) \in P(x)$, and $D_o(x,(u+v)/D_o(x,u) + D_o(x,v)) \leqq 1$, hence $D_o(x,u+v) \leqq D_o(x,u) + D_o(x,v)$, i.e., D_o is sub-additive in outputs. Finally take $u^\circ = \lambda u$, and $v^\circ = (1-\lambda)v$, then by homogeneity $D_o(x,\lambda u + (1-\lambda)v) \leqq \lambda D_o(x,u) + (1-\lambda)D_o(x,v)$.

(D_o.7 $\implies$ P.7) This follows directly from convexity of $D_o(x,u)$ in outputs.

(P.8 $\iff$ D_o.8) This proof is left to the reader.

$$Q.E.D.$$

A couple of comments on the above properties may be of interest. First we note that not every property on the output correspondence has an analog on the distance function: P.4 is the exception. Similarly, the output distance function is homogeneous in outputs, but there is no homogeneity condition imposed on the output correspondence. To elucidate this condition, let us define

(3.1.7)   **Definition:** The output correspondence is homogeneous of degree $+\alpha$ if
$$P(\lambda x) = \lambda^\alpha P(x), x \in \Re_+^N, P(x) \neq 0.$$

Homogeneity of degree α of the output correspondence is equivalent to homogeneity of degree $-\alpha$ in inputs of the output distance function. Thus the output distance function is always homogeneous in outputs, however it is homogeneous in inputs if and only if the correspondence is homogeneous of the negative degree. To prove our claim, suppose P is homogeneous of degree $+\alpha$, by the definition of D_o, for $\lambda > 0$,

$$
\begin{aligned}
D_o(\lambda x, u) &= \inf\{\theta > 0 : (u/\theta) \in P(\lambda x)\} \\
&= \inf\{\theta > 0 : (u/\theta) \in \lambda^\alpha P(x)\} \\
&= \lambda^{-\alpha} \inf\{(\lambda^\alpha \theta) > 0 : (u/\lambda^\alpha \theta) \in P(x)\} \\
&= \lambda^{-\alpha} D_o(x, u).
\end{aligned}
$$

The converse follows from the observation that $P(x) = \{u : D_o(x,u) \leqq 1\}$.

We turn next to the definition of the indirect distance function.

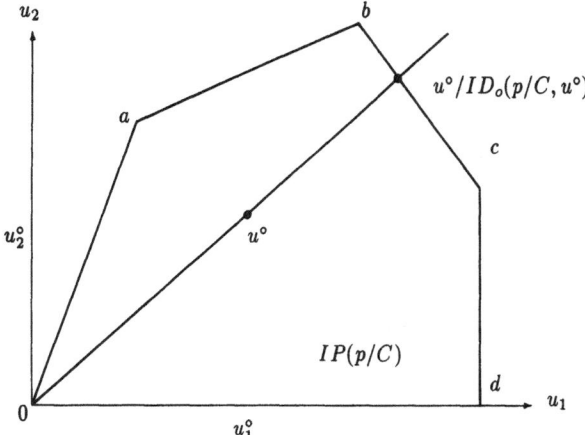

Figure 3.3: The Cost Indirect Output Distance Function

(3.1.8)   **Definition:** The function $ID_o : \Re_+^N \times \Re_+^M \longrightarrow \overline{\Re}_+$ defined by $ID_o(p/C, u) =$
inf$\{\theta > 0 : (u/\theta) \in IP(p/C)\}$ is called the *Cost Indirect Output Distance Function.*

Figure 3.3 illustrates the indirect distance function in the two output case. The indirect output
set is bounded by ($0abcd0$), and $u^\circ = (u_1^\circ, u_2^\circ)$ belongs to the set. The distance function expands
the output vector u° proportionally as much as possible while remaining feasible. In particular, it
is equivalent to

$$(3.1.9) \qquad ID_o(p/C, u^\circ) = \frac{\parallel u^\circ \parallel}{\parallel u^\circ/ID_o(p/C, u^\circ) \parallel}.$$

Figure 3.4 illustrates the distance function in input price, output quantity space. First, however,
we derive a simple cost indirect production function from the direct production function. Let
$F(x) = x^2, x \in \Re_+$, and $px \leqq C$. The indirect production function associated with $u = x^2$ equals
$u = 1/(p/C)^2$. Figure 3.4 illustrates. In our figure, $u^\circ \leqq IF((p/C)^\circ)$, hence
$IF((p/C)^\circ) = (1/(p/C)^\circ)^2$. Thus,

$$(3.1.10) \qquad ID_o((p/C)^\circ, u^\circ) = u^\circ/IF((p/C)^\circ).$$

In the single output case, the indirect output distance function equals output divided by the
indirect production function. This result also holds for $N > 1$.

41

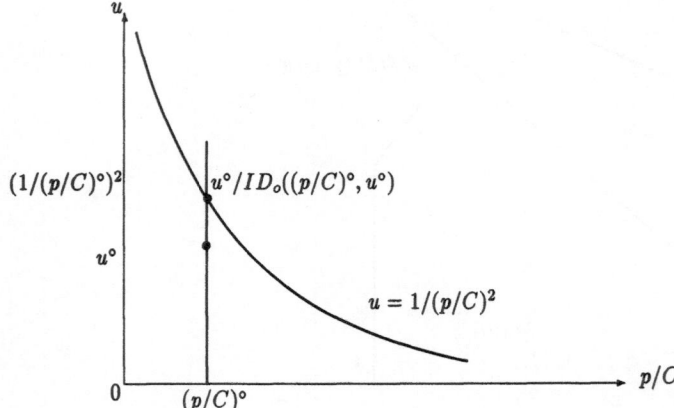

Figure 3.4: Example of The Indirect Production and Output Distance Functions

Whenever input prices are strictly positive, we showed that the indirect output correspondence may also be defined in terms of the cost function (Proposition (2.2.5)). Thus the corresponding distance function may in this case be expressed as

(3.1.11) $ID_o(p/C, u) = \inf\{\theta > 0 : C(u/\theta, p/C) \leq 1\}.$

(3.1.12)   **Proposition:** $ID_o(p/C, u) = C(u, p/C)$ for all $(p/C) > 0, u \in RangP$, if and only if the technology exhibits constant returns to scale (CRS), i.e.,

$C(\theta u, p/C) = \theta C(u, p/C), \theta > 0.$

At this point we take the condition $C(\theta u, p/C) = \theta C(u, p/C)$ to mean that the technology exhibits CRS, although the equivalence will be demonstrated. The proof of Proposition (3.1.12) below exploits the homogeneity of the indirect output distance function in outputs (the proof of homogeneity is straightforward and omitted).

Proof: ($\Longrightarrow$) Suppose $ID_o(p/C, u) = C(u, p/C)$, and that $\theta > 0$, then

(3.1.13) $C(\theta u, p/C) = ID_o(p/C, \theta u) = \theta ID_o(p/C, u) = \theta C(u, p/C).$

($\Longleftarrow$) Suppose $C(\theta u, p/C) = \theta C(u, p/C)$, then by (3.1.11) and (2.2.5),

$$
\begin{aligned}
ID_o(p/C, u) &= \inf\{\theta > 0 : C(u/\theta, p/C) \leq 1\} = \inf\{\theta > 0 : (1/\theta)C(u, p/C) \leq 1\} \\
&= \inf\{\theta : C(u, p/C) \leq \theta\} = C(u, p/C).
\end{aligned}
$$

42

Proposition (3.1.12) states that CRS is a necessary and sufficient condition for the indirect output distance function to equal the cost function. Our example in Figure 3.4 can be used to show that if a production function is homogeneous of degree $+2$, rather than of degree $+1$,

$ID_o(p/C, u) \neq C(u, p/C)$ for some $(p/C, u)$.

The discussion following the definition of the direct output distance function $D_o(x, u)$ also applies to show that Definition (3.1.8) of the indirect distance function is valid. Thus we may turn to the properties which $ID_o(p/C, u)$ inherits from the parent technology. We restrict ourselves to the case of positive prices so that $(p/C) > 0$. First, however, we need to show that the indirect output correspondence may be defined in terms of ID_o.

(3.1.14)   **Proposition** : $IP(p/C) = \{u : ID_o(p/C, u) \leqq 1\}$.

Proof: Suppose $u \in IP(p/C)$, then by the definition of $ID_o, ID_o(p/C, u) \leqq 1$. Conversely if $u \notin IP(p/C)$, then $ID_o(p/C, u) > 1$, and equality does not hold.

Proposition (3.1.14) shows that the indirect output distance function can be used to characterize the indirect output correspondence. Thus we have a function representation of $IP(p/C)$. Now to the properties.

   ID$_o$.1 $ID_o(p/C, 0) = 0$.

This condition states that when there is no output, and $(p/C) > 0$, the indirect distance function equals zero. To prove ID$_o$.1, note that $0 \in IP(p/C)$ for $(p/C) > 0$, see IP.1. Thus $ID_o(p/C, 0) = \inf\{\theta > 0 : (0/\theta) \in IP(p/C)\} = 0$. Conversely, to show that ID$_o$.1 implies IP.1, we note that $ID_o(p/C, u) = 0 \Longrightarrow (u/\theta) \in IP(p/C), \forall \theta > 0$ and the rest of the proof follows that used to show that P.1(a) $\Longleftrightarrow$ D$_o$.1(a).

   ID$_o$.2 $(p/C)^\circ \geqq (p/C) \Longrightarrow ID_o((p/C)^\circ, u) \geqq ID_o(p/C, u), u \in \Re_+^M$.

Monotonicity in cost deflated (positive) input prices is modeled by ID$_o$.2. The property is equivalent to IP.2. The proof is omitted since it is similar to the proof of P.2 $\Longleftrightarrow$ D$_o$.2. Disposability of outputs is modeled by the next two properties.

   ID$_o$.3 $ID_o(p/C, \theta u) \leqq ID_o(p/C, u), 0 \leqq \theta \leqq 1$.

43

ID$_o$.3.S $ID_o(p/C, u) \leqq ID_o(p/C, v), u \leqq v.$

The first property is weak disposability of outputs, and it is equivalent to IP.3. ID$_o$.3.S models strong disposability of outputs, and it is equivalent to IP.3.S. We prove the latter. Assume that ID.3.S holds and that $u \leqq v$. Then $ID_o(p/C, u) = \inf\{\theta > 0, (u/\theta) \in IP(p/C)\} = \inf\{\theta > 0 : C(u/\theta, p/C) \leqq 1\} \leqq \inf\{\theta > 0 : C(v/\theta, p/C) \leqq 1\} = ID_o(p/C, v)$. The reader should convince him (her) self about this proof, since we have used a property of the cost function which has not been proved. To prove the converse, one may apply Proposition (3.1.13).

ID$_o$.4 $ID_o(p/C, \theta u) = \theta ID_o(p/C, u), \theta > 0.$

Homogeneity of degree $+1$ in outputs is displayed by property ID$_o$.4. This property is a direct consequence of the definition of ID$_o$. The proof is omitted, but we note that ID$_o$.4 has no relation to returns to scale.

ID$_o$.4 (a) $ID_o(p/C, u)$ is lower semi-continuous on $\Re_+^M$, see A.6 in Appendix A; (b)
$ID_o(p/C, u)$ is lower semi-bounded on $\Re_1^M$, see A.14 in Appendix A.

To prove ID$_o$.5(a), we note that the cost function is lower semi-continuous in (u). Therefore, $\{u : ID_o(p/C, u) \leqq \alpha\}$ is closed for all $\alpha \geqq 0$, which proves ID$_o$.5(a). A similar argument applies to prove ID$_o$.5(b).

ID$_o$.6 If $ID_o(p/C, u) \leqq 1, u \geq 0$, then for all $\theta > 0$ there exists a λ_θ such that
$$ID_o(\lambda_\theta p/C, u) \leqq 1/\theta.$$

The interpretation of ID$_o$.6 is given after property IP.6, and is not repeated. The proof is similar to D$_o$.6 and is therefore omitted.

To sum up, we have proved that if the indirect output correspondence satisfies (IP.1-IP.6) then and only then does the indirect output distance function satisfy (ID$_o$.1 − ID$_o$.6). Observe of course, that we have assumed that prices are positive.

For empirical studies using distance functions, parametric forms may be required. Here we supply two examples. First, the *Translog Output Distance Function* is defined as

$$(3.1.15) \quad \ln D_o(x, u) = \alpha_o + \sum_{n=1}^{N} \alpha_n \ln x_n + \sum_{m=1}^{M} \beta_m \ln u_m + \frac{1}{2} \sum_{n=1}^{N} \sum_{n'=1}^{N} \alpha_{nn'}(\ln x_n)(\ln x_{n'})$$
$$+ \frac{1}{2} \sum_{m=1}^{M} \sum_{m'=1}^{M} \beta_{mm'}(\ln u_m)(\ln u_{m'}) + \sum_{n=1}^{N} \sum_{m=1}^{M} \gamma_{nm}(\ln x_n)(\ln u_m).$$

44

The parameters are restricted (1) in order that $D_o(x, u)$ be homogeneous of degree $+1$ in outputs by $\sum_{m=1}^{M} \beta_m = 1, \sum_{m\prime=1}^{M} \beta_{mm\prime} = \sum_{m=1}^{M} \gamma_{nm} = 0, m = 1, \cdots, M, n = 1, \cdots, N$ and (2) in order to impose symmetry by $\alpha_{nn\prime} = \alpha_{n\prime n}, \beta_{mm\prime} = \beta_{m\prime m}, n, n\prime = 1, \cdots, N$ and $m, m\prime = 1, \cdots, M$.

The *Cost Indirect Translog Output Distance Function* can be derived from (3.1.14) by merely substituting $(p_n/C), n = 1, \cdots, N$ for each $x_n, n = 1, \cdots, N$. We leave this to the reader and go on to introduce a *Variation of the Generalized Leontief Output Distance Function*, namely

$$(3.1.16) \qquad D_o(x, u) = \frac{\sum_{m=1}^{M} \sum_{m\prime=1}^{M} \beta_{mm\prime}(u_m u_{m\prime})^{\frac{1}{2}} + \sum_{n=1}^{N} \sum_{m=1}^{M} \gamma_{nm} x_n u_m}{\sum_{n=1}^{N} \sum_{n\prime=1}^{N} \alpha_{nn\prime}(x_n x_{n\prime})^{\frac{1}{2}}}.$$

This function is homogeneous of degree $+1$ in outputs (the reader should check this) and homogeneous of degree zero in its parameters. Due to homogeneity of degree zero we may normalize the parameters by

$$(3.1.17) \qquad \sum_{n=1}^{N} \sum_{n\prime=1}^{N} \alpha_{nn\prime} + \sum_{m=1}^{M} \sum_{m\prime=1}^{M} \beta_{mm\prime} + \sum_{n=1}^{N} \sum_{m=1}^{M} \gamma_{nm} = 1.$$

Other normalizations are, of course, possible. One may take $\alpha_{11} = 1$, for example. Note that since $D_o(x, u) \leqq 1$, for all feasible input output vectors, (3.1.16) can be transformed into a functional form which is linear in parameters, i.e.,

$$(3.1.18) \qquad \sum_{m=1}^{M} \sum_{m'=1}^{M} \beta_m \beta_{m\prime}(u_m u_{m\prime})^{\frac{1}{2}} + \sum_{n=1}^{N} \sum_{m=1}^{M} \gamma_{nm} x_n u_m - \sum_{n=1}^{N} \sum_{n\prime=1}^{N} \alpha_{nn\prime}(x_n x_{n\prime})^{\frac{1}{2}} \leqq 0.$$

Again by substituting $(p_n/C), n = 1, \cdots, N$ for $(x_n), n = 1, \cdots, N$, (3.1.16) may be transformed into a *Variation of the Cost Indirect Generalized Leontief Output Distance Function*. We leave this substitution to the reader.

The Translog function (3.1.14) has the drawback that it can not model technologies for which some inputs or outputs are zero. This problem is avoided in our function (3.1.16).

3.2 The Direct and Indirect Input Distance Functions

The input distance function models input sets as a function, in the same fashion as the output distance function models output sets as a function. It is defined by

$(3.2.1)$ **Definition:** The function $D_i : \Re_+^M \times \Re_+^N \longrightarrow \overline{\Re}_+$ defined by $D_i(u, x) =$ $\sup\{\lambda > 0 : (x/\lambda) \in L(u)\}$ is called the *(Direct) Input Distance Function*.

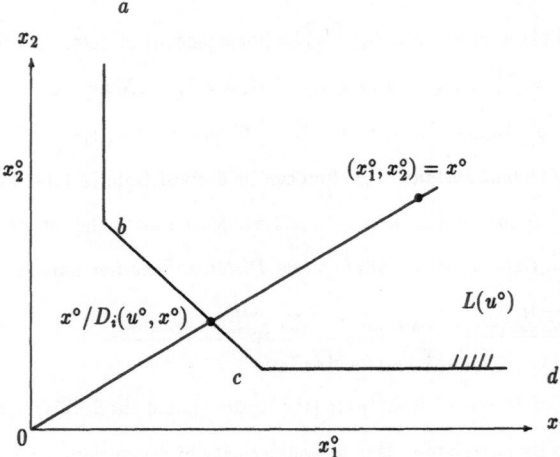

Figure 3.5: The Input Distance Function

Since $u \in P(x)$ if and only if $x \in L(u)$, we may also write the input distance function in terms of output sets, namely $D_i(u, x) = \sup\{\lambda > 0 : u \in P(x/\lambda)\}$. To illustrate Definition (3.2.1) suppose two inputs (x_1, x_2) are used to produce outputs, and suppose that $(x_1^\circ, x_2^\circ) \in L(u^\circ)$. Figure 3.5 illustrates. The input set $L(u^\circ)$ is bounded by $(abcd)$, and x° is feasible, i.e., $x^\circ \in L(u^\circ)$. The input distance function contracts x° proportionally as much as possible while remaining an element in $L(u^\circ)$, and it is equivalent to

$$(3.2.2) \qquad D_i(u^\circ, x^\circ) = \frac{\parallel x^\circ \parallel}{\parallel x^\circ / D_i(u^\circ, x^\circ) \parallel}.$$

In input output space the input distance function has the following interpretation. We assume that $M = N = 1$ and that the technology is represented by its graph, GR. The vector $(x^\circ, u^\circ) \in GR$, and since the input distance function takes output as exogenous, i.e., u° stays fixed, it contracts the input to x^1, where $x^1 = x^\circ / D_i(u^\circ, x^\circ)$. See Figure 3.6.

To convince ourselves that (3.2.1) is a valid definition, consider the four cases inverse to (P.a–P.d), namely

L.a $u \geq 0$ and $x \in L(u)$,

L.b $u \geq 0$, $x \notin L(u)$ but there exists a scalar λ such that $\lambda x \in L(u)$,

L.c $u \geq 0$ and $\lambda x \notin L(u)$ for any $\lambda > 0$,

46

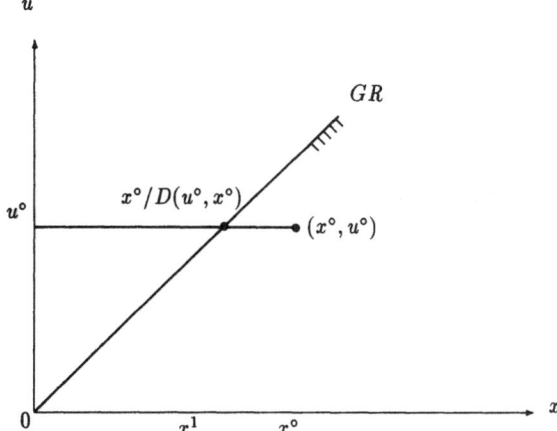

Figure 3.6: Simple Illustration of Input Distance Function

L.d $u = 0$.

In the first case, the supremum is a maximum, and $D_i(u, x) \geqq 1$. In case L.b a maximum is again obtained but $x \notin L(u)$ so $D_i(u, x) < 1$. Whenever $u \geq 0$ and there does not exist a scalar $\lambda > 0$ such that $\lambda x \in L(u)$, we take $D_i(u, x) = 0$. Since $L(0) = \Re_+^N$, $D_i(0, x)$ is taken to be $+\infty$.

Next, we show that the input distance function is a complete characterization of the input sets.

(3.2.3)   **Proposition** $L(u) = \{x : D_i(u, x) \geq 1\}, u \in \Re_+^M$.

Proof: Let $x \in L(u)$, if $u = 0, D_i(u, x) = +\infty$ and we are done, thus assume that $u \geq 0$ and $x \in L(u)$. In this case, $D_i(u, x) \geqq 1$. Conversely, if $u \geq 0$ but $x \notin L(u)$ then either $D_i(u, x) = 0$ or < 1, in both cases, $x \notin \{x : D_i(u, x) \geqq 1\}$.

Q.E.D.

The inverse relationship between $P(x)$ and $L(u)$, and the condition $P(x) = \{u : D_o(x, u) \leqq 1\}$ together with $L(u) = \{x : D_i(u, x) \geq 1\}$ imply that

(3.2.4) $D_o(x, u) \leqq 1 \Longleftrightarrow D_i(u, x) \geqq 1$.

It thus follows that $D_o(x, u) \leqq D_i(u, x)$ for all feasible input output vectors. To continue our study of the relationship between the two distance functions we may prove

(3.2.5)   **Proposition:** For all (x, u) such that $u \geq 0$ and $u \in P(x)$, $D_o(x, u) = (D_i(u, x))^{-1}$ if and only if P is homogeneous of degree $+1$.

47

Proof: Suppose that $u \geq 0, u \in P(x)$ and that $D_o(x, u) = (D_i(u, x))^{-1}$, then for $\lambda > 0$,

$$
\begin{aligned}
P(\lambda x) &= \{u : D_o(\lambda x, u) \leq 1\} = \{u : 1 \leq D_i(u, \lambda x)\} \\
&= \{u : 1 \leq \lambda D_i(u, x)\} = \{u : D_o(x, u/\lambda) \leq 1\} \\
&= \lambda \{v : D_o(x, v) \leq 1\} = \lambda P(x).
\end{aligned}
$$

To prove the converse, assume that P is homogeneous of degree $+1$ and that $u \in P(x)$ with $u \geq 0$. Then

$$
\begin{aligned}
D_o(x, u) &= \min\{\theta : (u/\theta) \in P(x)\} = \min\{\theta : u \in P(\theta x)\} \\
&= (\max\{\lambda : u \in P(x/\lambda)\})^{-1} = (\max\{\lambda : (x/\lambda) \in L(u)\})^{-1} \\
&= (D_i(u, x))^{-1}.
\end{aligned}
$$

$$Q.E.D.$$

Proposition (3.2.5) shows that a necessary and sufficient condition for the input and output distance functions to be reciprocals is constant returns to scale, or equivalently, that P is homogeneous of degree $+1$.

We can now list the properties which the input distance function inherits from the parent technology.

D$_i$.1   (a) $D_i(0, x) = +\infty, \forall x \in \Re_+^N$; (b) $D_i(u, 0) = 0, \forall u \geq 0$.

D$_i$.2 $\forall(u, x) \in \Re_+^{N+M}$ and $\lambda \geq 1, D_i(u, \lambda x) \geq D_i(u, x)$.

D$_i$.2.S $\forall u \in \Re_+^M$ and $x \geq y, D_i(u, x) \geq D_i(u, y)$.

D$_i$.3 $\forall(u, x) \in \Re_+^{M+N}, D_i(u, \lambda x) = \lambda D_i(u, x), \lambda > 0$.

D$_i$.4 $\forall(u, x) \in \Re_+^{M+N}, D_i(\theta u, x) \leq D_i(u, x), \theta \geq 1$.

D$_i$.4.S $\forall x \in \Re_+^N$ and $u \geq v, D_i(u, x) \leq D_i(v, x)$.

D$_i$.5   (a) If $(x^\ell, u^\ell) \longrightarrow (x^\circ, u^\circ)$ and $D_i(u^\ell, x^\ell) \geq 1$ for all ℓ, then $D_i(u^\circ, x^\circ) \geq 1$; (b) $D_i(u, x)$ is upper semi-continuous on $\Re_+^N$; (c) $D_i(u, x)$ is upper semi-bounded on $\Re_+^M$.

$D_i.6$ If $D_i(u, x) \geqq 1, u \geq 0$ and $x \geq 0$, then $\forall \theta \geqq 0$ there exists a scalar λ_θ such that $D_i(\theta u, x) \geqq 1/\lambda_\theta$.

$D_i.7$ $\forall x \in \Re_+^N, D_i(u, x)$ is quasi $-$ concave on $\Re_+^M$.

$D_i.8$ $\forall u \in \Re_+^M, D_i(u, x)$ is concave on $\Re_+^N$.

The following proposition may be proved.

(3.2.6) **Proposition:** The input distance function satisfies $(D_i.1 - D_i.8)$ if and only if the output correspondence satisfies $(P.1 - P.8)$.

The proof of this proposition is omitted. However, it is worth pointing out that it can be proved using either properties $(P.1 - P.8)$ or $(D_o.1 - D_o.8)$. In particular, if P is homogeneous of degree $+1$, then the relation between $(D_i.1 - D_i.8)$ and $(D_o.1 - D_o.8)$ is easily verified.

It was shown in Section 3.1 that the output correspondence is homogeneous of degree $+\alpha$ in inputs if and only if the output distance function is homogeneous of degree $-\alpha$ in inputs. In the same spirit one can prove the following

(3.2.7) **Proposition:** The input distance function is homogeneous of degree $-1/\alpha$ in outputs if and only if the output correspondence is homogeneous of degree $+\alpha$ in inputs.

Proof: Suppose that $P(\lambda x) = \lambda^\alpha P(x)$, then and only then is the output correspondence homogeneous of degree $-\alpha$ in output. Thus by (3.2.4), for $\theta > 0$,

$$
\begin{aligned}
D_i(\theta u, x) &= \max\{\lambda : D_o(x/\lambda, \theta u) \leqq 1\} = \max\{\lambda : \theta D_o(x/\lambda, u) \leqq 1\} \\
&= \max\{\lambda : D_o(\theta^{-\frac{1}{\alpha}} x/\lambda, u) \leqq 1\} = \theta^{-\frac{1}{\alpha}} D_i(u, x).
\end{aligned}
$$

The converse is proved similarly.

$$Q.E.D.$$

The last proposition shows that the output correspondence is homogeneous of degree $+\alpha$ in inputs if and only if the input distance function is homogeneous of degree $-1/\alpha$ in outputs, i.e., if and only if $D_o(x, u)$ is homogeneous of degree $+\alpha$ in input vectors.

We may now turn our attention to the revenue indirect input distance function, and we begin by defining it in terms of the revenue indirect input correspondence.

49

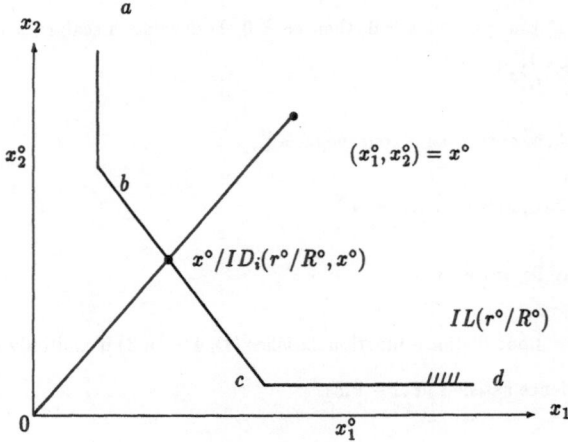

Figure 3.7: The Revenue Indirect Input Distance Function

(3.2.8)   **Definition:** The function $ID_i : \Re_+^M \times \Re_+^N \longrightarrow \overline{\Re}_+$ defined by $ID_i(r/R, x) =$ $\sup\{\lambda > 0 : (x/\lambda) \in IL(r/R)\}$ is called the *Revenue Indirect Input Distance Function.*

Figure 3.7 illustates the indirect input distance function in the case of two inputs. The indirect input set $IL(r^\circ/R^\circ)$ is bounded by $(abcd)$, the input vector $x^\circ = (x_1^\circ, x_2^\circ)$ is a member in the set. The indirect input distance function measures the distance from x° to the boundary of $IL(r^\circ/R^\circ)$, i.e.,

(3.2.9) $$ID_i(r^\circ/R^\circ, x^\circ) = \frac{\parallel x^\circ \parallel}{\parallel x^\circ/ID_i(r^\circ/R^\circ, x^\circ) \parallel}.$$

We may also illustrate this distance function in output price and input quantity space. To do this, assume that the production function may be written as $F(x) = x^2, x \in \Re_+$. The corresponding revenue function is $R = rx^2$. Figure 3.8 illustrates. In Figure 3.8, $r^\circ/R^\circ \geqq 1/(x^\circ)^2$, i.e., $(x^\circ, r^\circ/R^\circ)$ is "north" of the function $r/R = 1/x^2$, which denotes the boundary of the revenue indirect graph. The shape of the boundary follows from the increasing returns inherent in $F(x) = x^2$. The indirect distance function ID_o takes r°/R° as exogenous and contracts x° as much as possible, i.e., to $x^1 = x^\circ/ID_i(r^\circ/R^\circ, x^\circ)$.

Proposition (2.4.5) shows that we may express the indirect input correspondence via the revenue function. Therefore we may also express the indirect input distance function in terms of

50

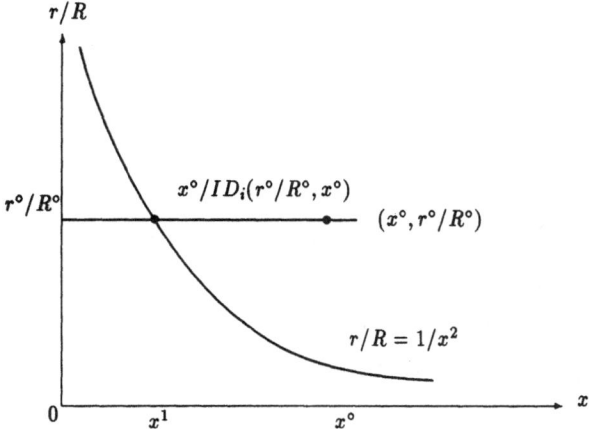

Figure 3.8: An Example of the Indirect Input Distance Function

revenue, i.e.,

$$(3.2.10) \quad ID_i(r/R, x) = \sup\{\lambda > 0 : R(x/\lambda, r/R) \leqq 1\},$$

and consequently we may ask under what conditions the two functions are equal. The answer is given in

(3.2.11)   **Proposition:** $ID_i(r/R, x) = R(x, r/R)$ for all $r/R \in \Re_+^M$ and $x \in \Re_+^N$ if and only if $R(\lambda x, r/R) = \lambda R(x, r/R), \lambda > 0.$

Proof: ($\Longrightarrow$) Suppose that $ID_i(r/R, x) = R(x, r/R)$, and that $\lambda > 0$, then by (3.2.10),

$$(3.2.12) \quad R(\lambda x, r/R) = ID_i(r/R, \lambda x) = \lambda ID_i(r/R, x) = \lambda R(x, r/R).$$

The second equality holds since the indirect input distance function is homogeneous of degree $+1$ in inputs.

($\Longleftarrow$) This follows from (3.2.10) and is left for the reader to complete.

$$Q.E.D.$$

Proposition (3.2.11) shows that the indirect input distance function equals the revenue function if and only if the revenue function is homogeneous of degree $+1$ in inputs. A sufficient condition for this homogeneity is homogeneity of the output correspondence. To prove the last claim suppose

51

$P(\lambda x) = \lambda P(x)$, then

$$\begin{aligned} R(\lambda x, r) &= \max_u \{ru : u \in P(\lambda x)\} = \max_u \{ru : u \in \lambda P(x)\} \\ &= \lambda \max\{r(u/\lambda) : (u/\lambda) \in P(x)\} = \lambda R(x, r). \end{aligned}$$

Recall that $ID_o(p/C, u) = C(u, p/C)$ if and only if the cost function is homogeneous of degree $+1$ in outputs. Thus, we would expect that Propositions (3.1.12) and (3.2.11) hold if and only if the technology exhibits constant returns to scale. To prove this statement requires a duality theorem which we have not yet introduced, therefore the proof is deferred.

Before stating the properties the indirect intput distance function inherits from the parent technology we prove

Proposition : $IL(r/R) = \{x : ID_i(r/R, x) \geqq 1\}$.

Proof: Suppose $x \in IL(r/R)$, then by the definition of $ID_i, ID_i(r/R, x) \geqq 1$. Conversely if $x \notin IL(r/R)$ then $ID_i(r/R, x) < 1$ and $x \notin \{x : ID_i(r/R, x) \geqq 1\}$.

$$Q.E.D.$$

The last proposition shows that the indirect input distance function characterizes the indirect input sets completely. Thus with the indirect input distance function we have a function that represents $IL(r/R)$. Next we turn to its properties.

$ID_i.1$ $ID_i(r/R, 0) = 0$.

$ID_i.2$ $(r/R)^\circ \geqq (r/R) \Longrightarrow ID_i((r/R)^\circ, x) \geqq ID_i(r/R, x), x \in \Re_+^N$.

$ID_i.3$ $ID_i(r/R, \lambda x) \geqq ID_i(r/R, x), \lambda \geqq 1$.

$ID_i.3.S$ $ID_i(r/R, x) \geqq ID_i(r/R, y), x \geqq y$.

$ID_i.4$ $ID_i(r/R, \lambda x) = \lambda ID_i(r/R, x), \lambda > 0$.

$ID_i.5$ $ID_i(r/R, x)$ is upper semi $-$ continuous on $\Re_+^N$.

$ID_i.6$ If $ID_i(r/R, x) \geqq 1, r/R \geq 0, x \geq 0$, then for all θ there exists a λ_θ such that $ID_i(\theta r/R, x) \geqq 1/\lambda_\theta$.

The relation between $(ID_i.1 - ID_i.6)$ and $(IL.1 - IL.6)$ is described by

52

(3.2.13)   **Proposition:** The revenue indirect input distance function satisfies $(ID_i.1 - ID_i.6)$ if and only if the indirect input correspondence satisfies $(IL.1 - IL.6)$.

Instead of proving Proposition (3.2.13) we turn our interest to two parametric formulations of the direct and the indirect input distance functions. First, the *Translog Input Distance Function* is specified as

$$(3.2.14) \quad lnD_i(u,x) = \alpha_o + \sum_{m=1}^{M} \alpha_m \ln u_m + \sum_{n=1}^{N} \beta_n ln x_n + \frac{1}{2} \sum_{m=1}^{M} \sum_{m\prime=1}^{M} \alpha_{mm\prime}(\ln u_m)(\ln u_{m\prime})$$
$$+ \frac{1}{2} \sum_{n=1}^{n} \sum_{n\prime=1}^{n} \beta_{nn\prime}(\ln x_n)(\ln x_{n\prime}) + \sum_{m=1}^{M} \sum_{n=1}^{N} \gamma_{mn}(\ln u_m)(\ln x_n).$$

The parameters of (3.2.14) are restricted first by $\sum_{n=1}^{N} \beta_n = 1, \sum_{n\prime=1}^{N} \alpha_{nn\prime} = \sum_{n=1}^{N} \gamma_{mn} = 0$, $m = 1, \cdots, M, n = 1, \cdots, N$ in order to make $D_i(u,x)$ homogeneous of degree $+1$ in inputs, and second by $\alpha_{mm\prime} = \alpha_{m\prime m}, \beta_{nn\prime} = \beta_{n\prime n}, n, n\prime = 1, \cdots, N$, and $m, m\prime = 1, \cdots, M$ in order for symmetry to be satisfied. Note that zero values of data are not admitted under the translog form.

The *Revenue Indirect Translog Input Distance Function* may be defined as (3.1.14) with the substitution for $u_m, m = 1, \cdots, M$ of $(r_m/R), m = 1, \cdots, m$, and $ID_i(r/R, x)$ for $D_i(u,x)$. This substitution is left to the reader.

Next we introduce a *Variation of the Generalized Leontief Input Distance Function* (which readily admits zeros in the data), namely

$$(3.2.15) \quad D_i(u,x) = \frac{\sum_{n=1}^{N} \sum_{n\prime=1}^{N} \beta_{nn\prime}(x_n x_{n\prime})^{\frac{1}{2}} + \sum_{m=1}^{M} \sum_{n=1}^{N} \gamma_{mn} u_m x_n}{\sum_{m=1}^{M} \sum_{m\prime=1}^{M} \alpha_{mm\prime}(u_m u_{m\prime})^{\frac{1}{2}}}.$$

This function satisfies homogeneity of degree $+1$ in inputs with no further restrictions, however, symmetry should be imposed. Also the parameters of this function should be normalized to prevent degeneracy, see the discussion of the output distance function version of (3.2.15). Finally, by substitution for u_m by $(r_m/R), m = 1, \cdots, M$ and $ID_i(r/R, x)$ for $D_i(u,x)$, we obtain a *Revenue Indirect Variation of the Generalized Leontief Input Distance Function*. It is left to the reader to undertake the substitution.

3.3   Piecewise Linear Models

The purpose of this section is to model technology in an activity analysis framework which is equivalent to a piecewise linear framework. Assume that there are $k = 1, \cdots, K$ activities that produce $m = 1, \cdots, M$ outputs by means of $n = 1, \cdots, N$ inputs. The input coefficients x_{kn} denote the quantity of input n used by the kth activity at unit intensity. The output coefficients u_{km} denote the mth output quantity of the kth activity at unity intensity. The coefficients are required to satisfy

(3.3.1) (i) $x_{kn} \geqq 0, u_{km} \geqq 0, k = 1, \cdots, K, n = 1, \cdots, N, m = 1, \cdots, M.$

(ii) $\sum_{k=1}^{K} x_{kn} > 0, n = 1, \cdots, N.$

(iii) $\sum_{n=1}^{N} x_{kn} > 0, k = 1, \cdots, K.$

(iv) $\sum_{k=1}^{K} u_{km} > 0, m = 1, \cdots, M.$

(v) $\sum_{m=1}^{M} u_{km} > 0, k = 1, \cdots, K.$

The requirements that x_{kn} and u_{km} are nonnegative merely establish that inputs and outputs are nonnegative quantities. Condition (ii) means that each input is required by at least one activity and (iii) means that each activity uses at least one input. The first condition on outputs (iv) means that each output is producible and (v) means that each activity produces at least one output.

To formulate the piecewise linear technology we need to introduce the intensity variables, which serve to construct technology from observed outputs and inputs. Denote the vector of intensity variables by $z = (z_1, \cdots, z_K)$ and assume at first that $z \in \Re_+^K$. Each z_k denotes the intensity at which activity k is undertaken. To illustrate, assume that there is one activity, which produces one output by means of one input. The graph of this technology is illustrated in Figure 3.9. The graph is constructed from (x°, u°) as $\{(x, u) : u \leqq zu^\circ, x \geqq zx^\circ, z \geqq 0\}$, i.e., the graph is the cone containing the input output vector (x°, u°). In the general case the graph is given by

(3.3.2) $GR = \{(x, u) : u_m \leqq \sum_{k=1}^{K} z_k u_{km}, m = 1, \cdots, M, \sum_{k=1}^{K} z_k x_{kn} \leqq x_n, n = 1, \cdots, N,$
$z_k \geqq 0, k = 1, \cdots, K\}.$

The corresponding output and input correspondences are written as

(3.3.3) $P(x) = \{u : u_m \leqq \sum_{k=1}^{K} z_k u_{km}, m = 1, \cdots, M, \sum_{k=1}^{K} z_k x_{kn} \leqq x_n, n = 1, \cdots, N,$
$z_k \geqq 0, k = 1, \cdots, K\}, x \in \Re_+^N,$

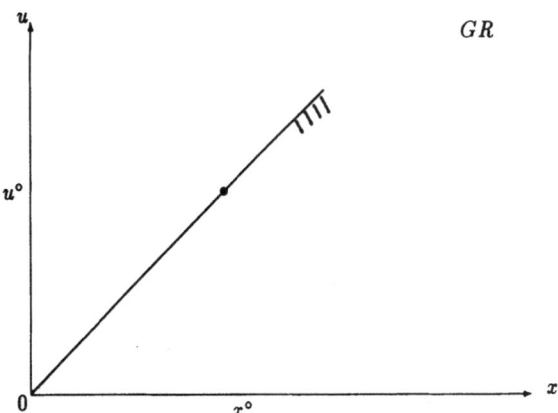

Figure 3.9: Piecewise Linear Model of GR under CRS

and

$$(3.3.4) \qquad L(u) = \{x : u_m \leqq \sum_{k=1}^{K} z_k u_{km}, m = 1, \cdots, M, \sum_{k=1}^{K} z_k x_{kn} \leqq x_n, n = 1, \cdots, N,$$
$$z_k \geqq 0, k = 1, \cdots, K\}, u \in \Re_{+}^{M}.$$

To illustrate $P(x)$, suppose there are two activities producing two outputs $(u_{11}, u_{12}), (u_{21}, u_{22})$ by means of one input. Also suppose that the two activities use the same amount of the input. Then Figure 3.10 illustrates the output set $P(x)$, where the output set is constructed from the convex combination of the two activities, i.e., the line segment (ab) including points a and b, as well as the combinations feasible given free disposbility of outputs, i.e., output combinations to the South of (ab) and to the West of (ab).

The properties that $P(x)$ satisfies are stated in

$(3.3.5)$ **Proposition:** The output correspondence $(3.3.3)$ is homogeneous of degree $+1$ and satisfies (P.1–P.9).

Proof: Let $\lambda > 0$, then

$$P(\lambda x) = \{u : u_m \leqq \sum_{k=1}^{K} z_k u_{km}, m = 1, \cdots, M, \sum_{k=1}^{K} z_k x_{kn} \leqq \lambda x_n,$$
$$n = 1, \cdots, N, z_k \geqq 0, k = 1, \cdots, K\}$$

55

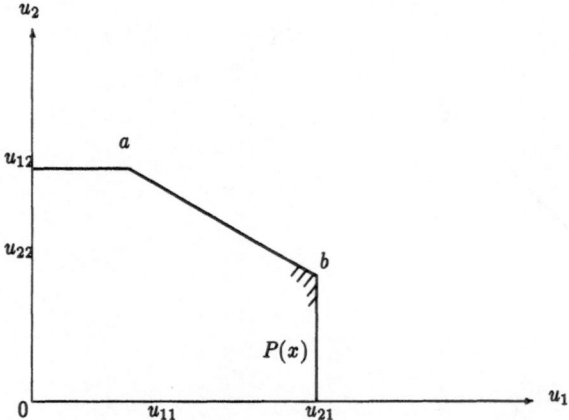

Figure 3.10: Piecewise Linear Output Set

$$
\begin{aligned}
&= \lambda\{(u/\lambda) : (u_m/\lambda) \leq \sum_{k=1}^{K}(z_k/\lambda)u_{km}, m = 1, \cdots, M, \\
&\qquad \sum_{k=1}^{K}(z_k/\lambda)x_{kn} \leq x_n, n = 1, \cdots, N, (z_k/\lambda) \geq 0, k = 1, \cdots, K\} \\
&= \lambda P(x).
\end{aligned}
$$

(P.1) Property (a) holds since $u \in \Re_+^M$ and $u_m \leq \sum_{k=1}^{K} z_k u_{km}, m = 1, \cdots, M$. The (b) property is valid since if $x = 0$, (ii) and (iii) imply that $z_k = 0, k = 1, \cdots, K$. Thus $u_m = 0, m = 1, \cdots, M$. Therefore $u \notin P(0)$ if $u \geq 0$.

(P.2.S) Because P.2.S $\implies$ P.2 we only prove that the piecewise linear model (3.3.3) satisfies strong disposability of inputs. Let $x \geq y$, then $Z(y) = \{(z_1, \cdots, z_K) : \sum_{k=1}^{K} z_k x_{kn} \leq y_n, n = 1, \cdots, N\}$ $\subseteq \{(z_1, \cdots, z_K) : \sum_{k=1}^{K} z_k x_{kn} \leq x_n, n = 1, \cdots, N\} = Z(x)$. Therefore, $\{u : u_m \leq \sum_{k=1}^{K} z_k u_{km},$ $m = 1, \cdots, M, z \in Z(y)\} \subseteq \{u : u_m \leq \sum_{k=1}^{K} z_k u_{km}, z \in Z(x)\}$ and $P(y) \subseteq P(x)$.

(P.3.S) Strong disposability of outputs follows from the inequalities $u_m \leq \sum_{k=1}^{K} z_k u_{km}, m = 1, \cdots, M$.

(P.4) The two conditions (ii) and (iii) on the input coefficients x_{kn} imply that for each $x \in \Re_+^N$, the set $Z(x) = \{(z_1, \cdots, z_k) : \sum_{k=1}^{K} z_k x_{kn} \leq x_n, n = 1, \cdots, N\}$ is bounded. Thus by the conditions of the output coefficients u_{km}, boundedness of $P(x)$ is implied.

(P.5) Suppose that $x^\ell \longrightarrow x^\circ, u^\ell \longrightarrow u^\circ$ and that $u^\ell \in P(x^\ell)$ for all ℓ. Then there exists a sequence $z^\ell = (z_1^\ell, \cdots, z_K^\ell)$ such that $u_m^\ell \leqq \sum_{k=1}^K z_k^\ell u_{km}, m = 1, \cdots, M$ and $\sum_{k=1}^K z_k^\ell x_{kn} \leqq x_n^\ell, n = 1, \cdots, N$ for all ℓ. We need to show that $z^\ell \longrightarrow z^\circ$ and that

$$u_m^\circ \leqq \sum_{k=1}^K z_k^\circ u_{km}, m = 1, \cdots, M, \sum_{k=1}^K z_k^\circ x_{kn} \leqq x_n^\circ,$$

$n = 1, \cdots, N$. Since the inequalities for both inputs and outputs are linear functions, we only need to prove that $z^\ell \longrightarrow z^\circ$. Therefore, let $\hat{x}_{n^\circ} = \sup_{\ell, n} \{x_n^\ell : n = 1, \cdots, N\}$. Since x^ℓ is convergent, $\hat{x}_{n^\circ}$ is bounded. Define the vector $\hat{x} = (\hat{x}_{n^\circ}, \cdots, \hat{x}_{n^\circ})$. By the conditions (ii) and (iii) on the input coefficients x_{kn}, the set $\{(z_1, \cdots, z_K) : \sum_{k=1}^K z_k x_{kn} \leqq \hat{x}_n, n = 1, \cdots, N\}$ is compact. Thus since this set includes the sequence z^ℓ, z^ℓ has a convergent subsequence $z^{\ell_k} \longrightarrow z^\circ$.

(P.6) This property follows from homogeneity.

The proofs of the last three properties are left to the reader.

$$Q.E.D.$$

If we restrict the intensity variables z_k to sum to one or less, the piecewise linear model becomes *Subhomogeneous*, i.e., for each $x \in \Re_+^N$ and $\lambda \geqq 1, P(\lambda x) \subseteqq \lambda P(x)$. To verify this claim, suppose that $x \in \Re_+^N$ and $\lambda \geqq 1$, then

$$
\begin{aligned}
P(\lambda x) \;=\; & \{u : u_m \leqq \sum_{k=1}^K z_k u_{km}, m = 1, \cdots, K, \sum_{k=1}^K z_k x_{kn} \leqq \lambda x_n, \\
& n = 1, \cdots, N, z_k \geqq 0, k = 1, \cdots, K, \sum_{k=1}^K z_k \leqq 1\} \\
\;=\; & \lambda\{(u/\lambda) : (u_m/\lambda) \leqq \sum_{k=1}^K (z_k/\lambda) u_{km}, m = 1, \cdots, M, \\
& \sum_{k=1}^K (z_k/\lambda) x_{kn} \leqq x_n, n = 1, \cdots, N, (z_k/\lambda) \geqq 0, k = 1, \cdots, K, (1/\lambda) \\
& \sum_{k=1}^K z_k \leqq 1/\lambda\} \\
\subseteqq\; & \lambda P(x)
\end{aligned}
$$

where the inclusion holds because $\lambda \geqq 1$. Note that the subhomogeneous piecewise linear technology does not meet axiom P.6, it does however satisfy (P.1-P.5) and (P.7, P.8). If the intensity variables are further restricted to sum to unity, a third piecewise linear model is obtained. This model allows for constant, increasing and decreasing returns to scale, but it does

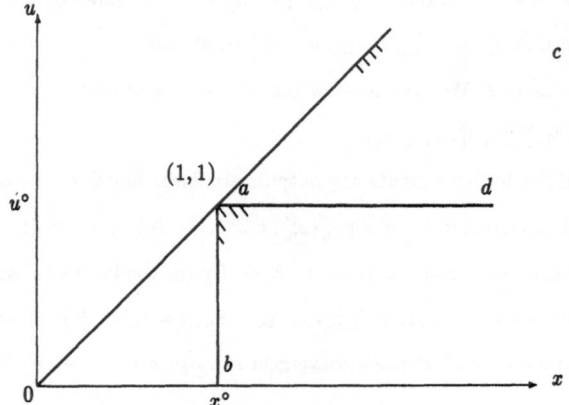

Figure 3.11: Piecewise Linear Models of GR

not satisfy P.1 or P.6. It may be written as

$$(3.3.6) \qquad P(x) = \{u : u_m \leqq \sum_{k=1}^{K} z_k u_{km}, m = 1, \cdots, M, \sum_{k=1}^{K} z_k x_{kn} \leqq x_n, n = 1, \cdots, N,$$

$$z_k \geqq 0, k = 1, \cdots, K, \sum_{k=1}^{K} z_k = 1\}, x \in \Re_+^N.$$

It is convenient to illustrate the three aforementioned piecewise linear models with their respective graphs, thus recall that $GR = \{(x, u) : u \in P(x), x \in \Re_+^N\}$. Suppose that there is just one activity, producing one output $u° = 1$, from one input $x° = 1$. The first model (3.3.3) is illustrated in Figure 3.11 by the cone bounded by the ray through a and the x-axis. The second piecewise linear model with the restriction $\sum_{k=1}^{K} z_k \leqq 1$ is bounded by $(0a)$ as well as the x-axis. The final model (3.3.6) is bounded by (ba) and the extension fron a through d and the x-axis starting at $x°$ going East.

So far both inputs and outputs have been treated as freely disposable. The next model is a variation of (3.3.3) with outputs strongly disposable but with inputs only weakly disposable.

$$(3.3.7) \qquad P(x) = \{u : u_m \leqq \sum_{k=1}^{K} z_k u_{km}, m = 1, \cdots, M, \sum_{k=1}^{K} z_k x_{kn} = \mu x_n,$$

$$n = 1, \cdots, K, 0 < \mu \leqq 1, z_k \geqq 0, k = 1, \cdots, K\}, x \in \Re_+^N.$$

Note that in (3.3.7) inputs are modeled by the equality $\sum_{k=1}^{K} z_k x_{kn} = \mu x_n$ with the scale factor μ. Compare this with (3.3.3) where inputs are modeled by the inequality $\sum_{k=1}^{K} z_k x_{kn} \leqq x_n$. The strict equality and scale factor in (3.3.7) allow for radial extensions of observed input bundles but not extensions due North or due East, i.e., (3.3.7) allows for backward bending isoquants, whereas (3.3.3) does not.

The final variation of (3.3.3) models the technology with weak disposability of outputs and strong disposability of inputs.

$$(3.3.8) \qquad P(x) \; = \; \{u : u_m = \sum_{k=1}^{K} \theta z_k u_{km}, m = 1, \cdots, M, \sum_{k=1}^{K} z_k x_{kn} \leqq x_n,$$
$$n = 1, \cdots, N, 0 \leqq \theta \leqq 1, z_k \geqq 0, k = 1, \cdots, K\}.$$

To prove that (3.3.8) satisfies weak disposability of outputs, let $u \in P(x)$, then there exist $z_k^o \geqq 0, k = 1, \cdots, K$, and θ^o such that $u_m = \sum_{k=1}^{K} \theta^o z_k^o u_{km}, m = 1, \cdots, M$. Now let $0 \leqq \mu \leqq 1$, then we need to show that $\mu u \in P(x)$, i.e., $\mu u \in \{u : u_m = \sum_{k=1}^{K} \theta^o z_k^o u_{km}, m = 1, \cdots, M, \sum_{k=1}^{K} z_k^o x_{kn} \leqq x_n, n = 1, \cdots, N, 0 \leqq \theta^o \leqq 1, z_k^o \geqq 0, k = 1, \cdots, K\}$. Since all outputs are scaled with μ, we may take $\theta' = \mu$ then $\mu u_m = \sum_{k=1}^{K} \theta' z_k^o u_{km}, m = 1, \cdots, M$. In Chapter 2, we showed that the cost indirect output correspondence could be expressed in two equivalent ways, one in terms of the cost function and one in terms of the direct output correspondence with a cost constraint. Here we adopt the latter formulation. Denote the input price vector by $p, p \in \Re_+^N$ and target cost by $C, C > 0$. The *Cost Indirect Piecewise Linear Output Correspondence* may then be written as

$$(3.3.9) \qquad IP(p/C) \; = \; \{u : u_m \leqq \sum_{k=1}^{K} z_k u_{km}, m = 1, \cdots, M,$$
$$\sum_{k=1}^{K} z_k x_{kn} \leqq x_n, n = 1, \cdots, N, z_k \geqq 0, k = 1, \cdots, K,$$
$$\sum_{n=1}^{N} p_n x_n \leqq C\}, (p/C) \in \Re_+^N.$$

The output sets in (3.3.9) are defined for $(p/C) \in \Re_+^N$, i.e., we allow some or all prices to be zero. In view of Proposition (2.2.5) our two definitions of $IP(p/C)$ may then not coincide. However, the efficient input set $EffL(u)$, see (2.A.3) is bounded (prove this) therefore the cost function can be defined as a minimum, see (2.A.7), and hence we can show that Proposition (2.2.5) holds even when $p \geqq 0$.

(3.3.10)   **Proposition:** The cost indirect output correspondence (3.3.9) is homogeneous of degree -1 and satisfies (IP.1–IP.7).

Proof: Suppose that $\lambda > 0$, then

$$
\begin{aligned}
IP(\lambda p/C) &= \{u : u_m \leqq \sum_{k=1}^{K} z_k u_{km}, m = 1, \cdots, M, \sum_{k=1}^{K} z_k x_{kn} \leqq x_n, \\
&\qquad n = 1, \cdots, N, z_k \geqq 0, k = 1, \cdots, K, \sum_{n=1}^{N} (\lambda p_n) x_n \leqq C\} \\
&= (1/\lambda)\{\lambda u : \lambda u_m \leqq \sum_{k=1}^{K} (\lambda z_k) u_{km}, m = 1, \cdots, M, \\
&\qquad \sum_{k=1}^{K} (\lambda z_k) x_{kn} \leqq \lambda x_n, n = 1, \cdots, N, \\
&\qquad (\lambda z_k) \geqq 0, k = 1, \cdots, K, \sum_{n=1}^{N} p_n(\lambda x_n) \leqq C\} \\
&= (1/\lambda)\{v : v_m \leqq \sum_{k=1}^{K} \xi_k u_{km}, m = 1, \cdots, M, \\
&\qquad \sum_{k=1}^{K} \xi_k x_{kn} \leqq y_n, n = 1, \cdots, N, \xi_n \geqq 0, k = 1, \cdots, K, \sum_{n=1}^{N} p_n y_n \leqq C\} \\
&= (1/\lambda) IP(p/C)
\end{aligned}
$$

(IP.1) Part (a) follows from the inequalities $u_m \leqq \sum_{k=1}^{K} z_k u_{km}, m = 1, \cdots, M$. If $p = 0$, then clearly the sum $\sum_{n=1}^{N} p_n x_n \leqq C$, does not add any effective constraints to the model. Therefore $IP(0) = RangP$, which proves that part (b) holds.

(IP.2) Suppose that $(p/C)^\circ \geqq (p/C)$ then the budget set $B((p/C)^\circ) \subseteqq B(p/C)$ and therefore $IP((p/C)^\circ) \subseteqq IP(p/C)$.

(IP.3.S) This property is a consequence of the inequalities $u_m \leqq \sum_{k=1}^{K} z_k u_{km}, m = 1, \cdots, M$.

(IP.4) Whenever $p > 0$, the budget set $B(p/C)$ is bounded and therefore the set $Z(p/C) = \{(z_1, \cdots, z_K) : \sum_{k=1}^{K} z_k x_{kn} \leqq x_n, n = 1, \cdots, N, z_k \geqq 0, k = 1, \cdots, K, \sum_{n=1}^{N} p_n x_n \leqq C\}$ is bounded because (ii) and (iii) of (3.3.1) hold. Finally by (iv) and (v) and the boundedness of $Z(p/C)$, the output set $IP(p/C)$ is bounded.

(IP.5) The proof of this property is similar to the proof that the direct piecewise linear output correspondence has a closed graph, thus we leave it to the reader.

We also leave the proof of IP.6 and IP.7 to the reader.

$$Q.E.D.$$

Since the indirect output correspondence (3.3.9) is homogeneous of degree -1, it may be written in the form

(3.3.11) $IP(p/C) = C\ IP(p)$.

From the discussion of the direct output correspondence it is clear that we may modify (3.3.9) to satisfy various forms of returns to scale, by imposing various constraints on the intensity variables $z_k, k = 1, \cdots, K$. Similarly, we can model weak disposability of outputs by varying the output constraints. We leave these modifications to the reader, and go on to introduce the *Revenue Indirect Piecewise Linear Input Correspondence*.

$$(3.3.12)\quad IL(r/R) = \{x : u_m \leq \sum_{k=1}^{K} z_k u_{km}, m = 1, \cdots, M, \sum_{k=1}^{K} z_k x_{kn} \leq x_n, n = 1, \cdots, N,$$
$$z_k \geq 0, k = 1, \cdots, K, \sum_{m=1}^{M} r_m u_m \geq R\}, (r/R) \in \Re_+^M.$$

This correspondence is homogeneous of degree -1 in output prices, therefore we may also write it as

(3.3.13) $IL(r/R) = R \cdot IL(r)$,

where $IL(r)$ can be written as in (3.3.12) with $R = 1$. If we ignore the revenue constraint $\sum_{m=1}^{M} r_m u_m \geq R$, in (3.3.12), and assume that outputs are taken as given, then (3.3.12) becomes the input correspondence (3.3.4). One may also prove

(3.3.14)   **Proposition**: The revenue indirect input correspondence (3.3.12) is homogeneous of degree -1 and satisfies (IL.1–IL.7).

These proofs are left to the reader.

3.P Problems

(3.P.1)   Suppose that $x \in \Re_+^N, u \in \Re_+$. If $F(x)$ is a production function show that the output distance function can be written as $D_o(x, u) = u/F(x)$.

(3.P.2)   Suppose that $P(x) \neq \{0\}$. Prove that if $P(x) = \{u : D_o(x, u) \leq 1\}$, then outputs are weakly disposable.

(3.P.3) Prove the equivalence: P.2.S $\iff$ D_o.2.S.

(3.P.4) Prove the equivalence: P.3.S $\iff$ D_o.4.S.

(3.P.5) Prove that the output correspondence is homogeneous of degree +1 if and only if the input correspondence is homogeneous of degree +1.

(3.P.6) Suppose that the output correspondence is homogeneous of degree +1. Prove the equivalence $(D_i.1 - D_i.8)$ $\iff$ $(D_o.1 - D_o.8)$.

(3.P.7) Show that the output correspondence is homogeneous of degree α in inputs if and only if the input correspondence is homogeneous of degree $1/\alpha$ in outputs.

(3.P.8) Complete the proof of (3.2.1).

(3.P.9) Prove property P.7 of Proposition (3.3.5).

(3.P.10) Prove that the output correspondence (3.3.7) satisfies weak disposability of inputs.

(3.P.11) Prove that the indirect output correspondence (3.3.9) meets IP.5.

(3.P.12) Prove Proposition (3.3.14).

Notes

The direct and indirect distance function representations of the technologies discussed in this chapter are based on Shephard (1970), Shephard (1974), and Shephard and Färe (1980). The piecewise linear model goes back to von Neumann (1938; 1945) with additional contributions by Koopmans (1951), Afriat (1972), and Shephard (1974). Other piecewise models have been employed including piecewise Cobb-Douglas (see Charnes, Cooper, Sieford and Stutz (1983)) which is generalized in Färe, Grosskopf, Njinkeu (1988). The indirect piecewise linear model was introduced and used to measure performance in school districts in Färe, Grosskopf and Lovell (1988).

Chapter 4

Indirect Revenue Maximization, Indirect Cost Minimization and Profit Maximization

4.0 Introduction

Chapters 1, 2 and 3 have established a variety of representations of technology, with and without constraint. In this chapter we employ these to model various optimization problems typically faced by firms, modified to include budget and revenue targets.

We begin with cost indirect revenue maximization which yields the maximum value of output from a given budget. It can be defined in terms of the indirect output distance function or cost function. We summarize the properties which are inherited from the parent technology.

These optimization problems are, of course, closely related to the standard (direct) problems and can also be modeled as constrained maximization (minimization) problems using standard Lagrange multiplier formulation. In this framework we can derive the derivative properties of the indirect revenue function. This suggests the possibility of uncovering Shephard type lemmas in the indirect framework. After demonstrating the duality between the indirect revenue function and the indirect output distance function, we summarize these envelope properties of the indirect revenue and output distance functions, which can be used to solve for shadow prices and quantities. We also derive what we call the cost indirect conjugate duality theorem.

In order to provide a framework for exploiting these derivative and duality results, we include two parameterizations of the indirect revenue function. In contrast to the direct revenue function which depends on output prices and input quantities, the indirect revenue function depends only on prices.

Following our discussion of indirect revenue maximization is a parallel development of indirect cost minimization. As in the revenue case, we derive the properties the indirect cost function inherits from the parent technology, illustrate its formulation as a Lagrangian problem, demonstrate the duality between indirect cost and revenue functions, summarize the derivative properties and provide two parameterizations of the revenue indirect cost function.

A short section on profit maximization is included in Section 4.3. The purpose of this section is to provide concrete examples using simple explicit technologies. Three variations on the profit maximization theme are included: expenditure constrained profit maximization, cost indirect profit maximization and revenue indirect profit maximization. These are intended to illustrate how to calculate these indirect functions as well as illustrating the duality relationships among them.

An appendix includes the dualities between: (1) cost and input distance functions and (2) revenue and output distance functions. These are exploited in the summary of derivative properties of the cost and revenue functions, including of course, Shephard's lemma.

A set of problems and notes to the literature concludes the chapter.

4.1   Revenue Maximization

The cost indirect output correspondence was represented by the indirect output distance function in Section 3.1. The dual representation is introduced in this section. In particular, we define the cost indirect revenue function as the maximal output value that can be achieved when input and output prices are known, and the budget is constrained not to exceed a target. We prove that under some regularity conditions, in particular the convexity of the indirect output sets, the indirect output correspondence may be represented by either the indirect revenue function or the indirect output distance function.

Next we consider the properties which the indirect revenue function inherits from the parent distance function, some of which are similar to the properties of the (direct) revenue function. To explore these properties, suppose output prices are represented by a nonnegative vector

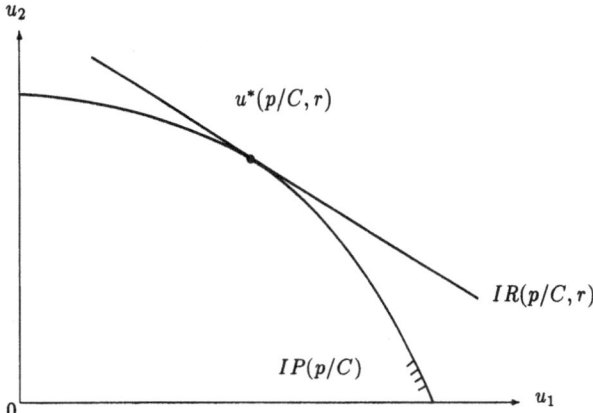

Figure 4.1: The Cost Indirect Revenue Function

$r = (r_1, \cdots, r_M)$. The *Cost Indirect Revenue Function* is defined for $(p/C) \in \Re_{++}^N$ by

(4.1.1) $IR(p/C, r) = \max_{u}\{ru : u \in IP(p/C)\} = \max_{u}\{ru : u \in P(x), px \leqq C\}.$

Suppose two outputs are produced, and that $IP(p/C) \neq 0$, then Figure 4.1 illustrates Definition (4.1.1). The output prices (r_1, r_2) are positive, and the indirect output set is given by $IP(p/C)$. The maximizing output vector is denoted $u^*(p/C, r)$, and $IR(p/C, r) = ru^*(p/C, r)$.

The assumption that cost deflated input prices are positive implies that the output set $IP(p/C)$ is compact, see Section 2.2. Thus since $ru = \sum_{m=1}^{M} r_m u_m$ is continuous and $IP(p/C) \neq \emptyset$ by IP.1, the maximum in (4.1.1) is achieved, and the definition is valid. The indirect output correspondence $IP(p/C)$, may be represented by a distance function, therefore we may also write

(4.1.2) $IR(p/C, r) = \max_{u}\{ru : ID_o(p/C, u) \leqq 1\},$

where the distance function $ID_o(p/C, u)$ is defined by (3.1.8). Moreover, since cost deflated input prices are positive, the indirect revenue function can be defined in terms of the cost function $C(u, p)$ by

(4.1.3) $IR(p/C, r) = \max_{u}\{ru : C(u, p) \leqq C\},$

65

see Section 2.2. Suppose that the cost function and the indirect revenue function are differentiable, we may then obtain a first set of derivative properties based on the Lagrangian formulation below.

(4.1.4) $IR(p/C, r) = ru + \theta(C(u, p/C) - 1).$

The derivative properties are

(4.1.5) $\nabla_r IR(p/C, r) = u(p/C, r),$

(4.1.6) $\nabla_{p/C} IR(p/C, r) = \theta \nabla_{p/C} C(u, p/C).$

The first property yields the vector of cost indirect supply functions. Each supply function $u_m(p/C, r)$ depends on cost deflated input prices $(p/C) = (p_1/C, \cdots, p_N/C)$ and output prices $r = (r_1, \cdots, r_M)$. In order to interpret the second property, we note that if Shephard's lemma holds, see Appendix, then the gradient with respect to p/C yields the cost deflated factor demands $x(u, p/C) = \nabla_{p/C} C(u, p/C)$. The interpretation of the Lagrangian multiplier θ is obtained from the first order conditions associated with expression (4.1.4). These conditions are

(4.1.7)   (a) $r_m = \theta \dfrac{\partial C(u, p/C)}{\partial u_m}, m = 1, \cdots, M,$

            (b) $C(u, p/C) = 1.$

By multiplying expression (a) by u_m, summing over $m = 1, \cdots, M$, we arrive at an expression for scale elasticity based on the cost function $\varepsilon_c(u, p/C) = \left[\sum_{m=1}^{M} u_m \frac{\partial C(u, p/C)}{\partial u_m} \right]^{-1}$ where $(C(u, p/C) = 1$ by (b)). Using this elasticity, substituting into (4.1.7) and rearranging yields $\theta = IR\varepsilon_c$. The expression (4.1.6) can now be written as

(4.1.8) $\nabla_{p/C} IR(p/C, r) = x IR\varepsilon_c.$

To be "absolutely" precise, we must observe that outputs $u_m, m = 1, \cdots, M$ are functions of (p/C) and r, thus the input demands $x_n, n = 1, \cdots, N$, are also functions of (p/C) and r. Therefore, the right hand side of (4.1.8) is, as it must be, a function of cost deflated input prices (p/C) and output prices r. The factor demand equations in (4.1.8), i.e.,

(4.1.9) $x_n(p/C, r) = \dfrac{\nabla_{p_n/C} IR(p/C, r)}{\varepsilon_c IR(p/C, r)}, n = 1, \cdots, N,$

reduce to $x_n(p/C,r) = \nabla_{p_n/C} IR(p/C,r)/IR(p/C,r)$ whenever the technology exhibits constant returns to scale, i.e., whenever $\varepsilon_c = 1$.

Note that factor demands derived from Shephard's lemma, $x(u,p)$, depend on output quantities and input prices. The factor demands derived from the indirect revenue function, $x(p/C,r)$, depend on cost-deflated input prices and output prices. Thus if the researcher is interested in the response of input demands to output prices, the input demands $x(p/C,r)$ can easily provide that information.

Suppose that $(p/C) \in \Re_{++}^N$ and that $r \in \Re_{+}^M$, since $IP(p/C) \neq \emptyset$ and compact, ru achieves a maximum on $IP(p/C)$. Therefore if IP.3.S holds, in addition to IP.1, IP.4 and IP.5, the indirect revenue function is

IR.1   nonnegative and nondecreasing in output prices,

IR.2   homogeneous of degree $+1$ in output prices,

IR.3   convex and continuous in (positive) output prices.

The proofs of the first two properties are left to the reader. The proof that a convex function is continuous on an open set, i.e., on $\Re_{++}^M$ can be found in Rockafeller (1970, p. 82). Thus we only prove that $IR(p/C,r)$ is convex in output prices. Suppose that $r^\circ, r' \in \Re_{+}^M$ and that $0 \leqq \theta \leqq 1$. Then

$$IR(p/C,r^\circ) = r^\circ u^\circ \text{ and } IR(p/C,r') = r'u', \text{ where } u^\circ, u' \in IP(p/C).$$

Moreover,

$$
\begin{aligned}
IR(p/C, \theta r^\circ + (1-\theta)r') &= (\theta r^\circ + (1-\theta)r')u^*(\theta), \ \ u^*(\theta) \in IP(p/C) \\
&\leqq \theta r^\circ u^\circ + (1-\theta)r'u' \\
&= \theta IR(p/C,r^\circ) + (1-\theta)IR(p/C,r'),
\end{aligned}
$$

this proves that $IR(p/C,r)$ is convex in nonnegative output prices.

Next, we prove that the indirect revenue function can be used to retrieve the indirect output sets $IP(p/C)$. Thus introduce the set

$$(4.1.10) \quad IP^*(p/C) = \{u : ru \leqq IR(p/C,r), \forall r \geq 0\} = \bigcap_{r \geq 0}\{u : ru \leqq IR(p/C,r)\}.$$

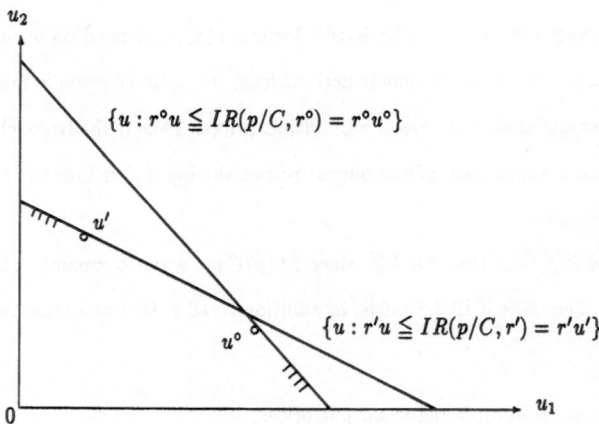

u_2

$\{u : r^\circ u \leqq IR(p/C, r^\circ) = r^\circ u^\circ\}$

u'

u°

$\{u : r'u \leqq IR(p/C, r') = r'u'\}$

u_1

0

Figure 4.2:

To illustrate the set $IP^*(p/C)$, suppose that there are two distinct output price vectors $r^\circ = (r_1^\circ, r_2^\circ)$ and $r' = (r_1', r_2')$. Figure 4.2 illustrates the lower halfspaces spanned by r° and r' together with their intersection. This intersection is part of the construction of an indirect output set $IP^*(p/C)$. Since the output set $IP^*(p/C)$ is obtained from the indirect revenue function and consists of the output vectors that belong to the intersection of all lower halfspaces determined by the revenue function, clearly $IP(p/C) \subseteqq IP^*(p/C)$. The following proposition shows when equality holds.

(4.1.11)　　**Proposition:** Suppose that the cost indirect output correspondence satisfies (IP.1, IP.3.S, IP.4 and IP.5) and that it has convex output sets $IP(p/C)$, then
$$\forall (p/C) \in \Re_{++}^N, \; IP(p/C) = IP^*(p/C).$$

Proof: Let $(p/C) \in \Re_{++}^N$, and suppose that $u^\circ \in IP(p/C)$. Then $ru^\circ \leqq IR(p/C, r)$ for all $r \geq 0$. Hence, $u^\circ \in IP^*(p/C)$. Conversely, assume that $u^\circ \in \Re_+^M$ but $u^\circ \notin IP(p/C)$. Since $IP(p/C)$ is compact, by the strict separation theorem (Rockafeller, 1970, p. 99) there exists an output price vector $r^\circ \in \Re^M, r^\circ \neq 0$ such that $r^\circ u^\circ > \max\{r^\circ u : u \in IP(p/C)\}$. Strong disposability of outputs, property IP.3.S, implies that $r^\circ \geq 0$. Thus $u^\circ \notin IP(p/C)$, and therefore, $IP(p/C) = IP^*(p/C)$.

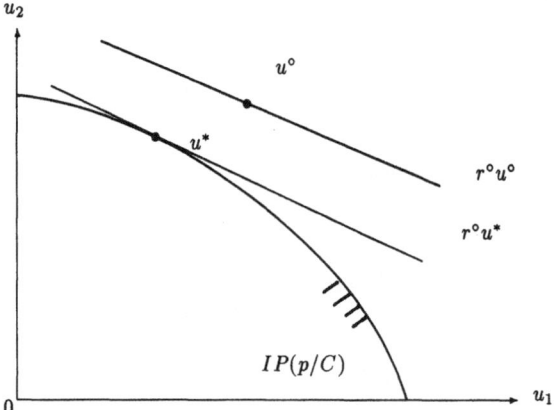

u_2

u°

u^*

$r^\circ u^\circ$

$r^\circ u^*$

$IP(p/C)$

0

u_1

Figure 4.3:

Q.E.D.

The thrust of Proposition (4.1.11) is illustrated in Figure 4.3. The output vector u° does not belong to $IP(p/C)$, and there is a price vector r° such that $r^\circ u^\circ > r^\circ u^*$, where u^* is the maximizer in $IP(p/C)$. Since $IP(p/C)$ is contained in $\{u : r^\circ u \leqq IR(p/C, r^\circ)\}$, u° cannot be an element of $IP^*(p/C)$.

Proposition (4.1.11) is the basic tool needed to prove that the cost indirect distance function is dual to the indirect revenue function, i.e.,

(4.1.12) $ID_o(p/C, u) = \max\limits_r \{ru : IR(p/C, r) \leqq 1\}.$

The proof of (4.1.12) also requires that

(4.1.13) $\{u : ru \leqq IR(p/C, r), \forall r \geq 0\} = \{u : ru \leqq 1, IR(p/C, r) \leqq 1\}.$

To prove this claim, first assume that $IR(p/C, r^\circ) = 0$ for some $r^\circ \geq 0$. Then $r^\circ u = 0$ and $\{u : r^\circ u \leqq IR(p/C, r^\circ)\} \subseteqq \{u : r^\circ u \leqq 1, IR(p/C, r^\circ) \leqq 1\}$. Moreover, since $IR(p/C, r^\circ)$ is the maximum, $IR(p/C, r^\circ) \geqq r^\circ u$. Thus equality holds. To continue, suppose that $u^\circ \notin \{u : ru \leqq IR(p/C, r), \forall r \geq 0\}$. Then, there exists a $r^\circ \geq 0$ such that $r^\circ u^\circ > IR(p/C, r^\circ)$, and we may suppose that $IR(p/C, r^\circ) > 0$. Thus $\frac{r^\circ u^\circ}{IR(p/C, r^\circ)} > 1$. Define $r' = r^\circ / IR(p/C, r^\circ)$, then

$r'u^\circ > 1$ and $IR(p/C, r') = IR(p/C, r^\circ/IR(p/C, r^\circ)) = 1$, hence

$u^\circ \notin \{u : ru \leqq 1, IR(p/C, r) \leqq 1\}$. To prove the converse, suppose that $u^\circ \in \{u : ru \leqq IR(p/C, r), \forall r \geq 0\}$. Then $ru^\circ \leqq IR(p/C, r), \forall r \geq 0$. Therefore, $r^\circ u^\circ \leqq IR(p/C, r^\circ) = 1, \forall r^\circ \geq 0$, where $r^\circ = r/IR(p/C, r)$, and $u^\circ \in \{u : ru \leqq 1, IR(p/C, r) \leqq 1\}$.

Q.E.D.

We may now prove that expression (4.1.12) holds. Thus let $(p/C) \in \Re_{++}^N, u \in Rang P$, then

$$
\begin{aligned}
ID_o(p/C, u) &= \min\{\theta : (u/\theta) \in IP(p/C)\} \\
&= \min\{\theta : (u/\theta) \in IP^*(p/C)\} (\text{by } (4.1.11)) \\
&= \min\{\theta : \frac{ru}{\theta} \leqq IR(p/C, r), \forall r \geq 0\} (\text{by } (4.1.10)) \\
&= \min\{\theta : ru \leqq \theta, IR(p/C, r) \leqq 1\} (\text{by } (4.1.13)) \\
&= \max_r\{ru : IR(p/C, r) \leqq 1\}.
\end{aligned}
$$

Q.E.D.

Expressions (4.1.2) and (4.1.12) form our first duality pair, and it is useful to have them stated together.

$$
\begin{aligned}
(4.1.14) \qquad IR(p/C, r) &= \max_u\{ru : ID_o(p/C, u) \leqq 1\} \\
ID_o(p/C, u) &= \max_r\{ru : IR(p/C, r) \leqq 1\}.
\end{aligned}
$$

Some observations are in place. First we note that the indirect revenue function is computed as the maximum revenue with respect to output quantities, with the indirect output distance function as the constraint. Secondly, the indirect output distance function is computed as the maximum revenue with respect to output prices, taking the indirect revenue function as the constraint. From a technical viewpoint, we have imposed convexity on the output sets $IP(p/C)$ and assumed strong disposability of outputs. Convexity must be maintained if equality in the second part of the duality pair is to hold. If convexity is not imposed, then the inequality, $ID_o(p/C, u) \geqq \max_r\{ru : IR(p/C, r) \leqq 1\}$ applies. Regarding disposability of outputs, we may relax our assumption to weak disposability and equality will still hold for both parts of (4.1.14). However, we must then allow for nonpositive prices, rather than $r \geq 0$. We explore the implications of nonpositive prices in our discussion of shadow pricing "bads", see Section 5.2. Finally, recall that throughout our discussion of duality, we have assumed that cost deflated input prices are positive.

We continue by determining the properties which $IR(p/C, r)$ inherits from the indirect output correspondence.

IR.4 $IR(p/C, 0) = 0, (p/C) \in \Re_{++}^N$

This property merely states that if the prices of outputs are all zero, total revenue is zero independent of how much you produce.

IR.5 $IR(0, r) = +\infty, r \geq 0.$

If input prices are zero, then there is no effective budget constraint and by IP.1(b), $IP(0) = RangP = \Re_+^M$, thus since some output prices are nonzero, there does not exist any bound on the production value. Note that in the case of $(p/C) = 0$, indirect revenue must be defined by a supremum rather than as a maximum.

IR.6 If $(p/C)^\circ \geq (p/C) > 0$, and $r \geq 0$ then $IR(p/C, r) \geq IR((p/C)^\circ, r)$.

Since the indirect output correspondence is nonincreasing in cost-deflated input prices, IP.2, the indirect revenue function, is nonincreasing in cost-deflated input prices as well.

Our cost indirect duality theorem (4.1.14) ensures that the indirect output distance function can be calculated from the indirect revenue function. Thus since the indirect output correspondence can be defined via the indirect output distance function, as $IP(p/C) = \{u : ID_o(p/C, u) \leq 1\}$, see (3.1.13), the properties on the output correspondence IP can be derived from those of the indirect revenue function. Also, of course, the properties on $ID_o(p/C, u)$ are dual to those on $IR(p/C, r)$.

Some additional derivative properties can be deduced from the duality theorem (4.1.14). The first part of that theorem may be expressed by

(4.1.15) $IR(p/C, r) = ru + \theta(ID_o(p/C, u) - 1).$

In order to interpret the Lagrangian θ, consider the problem

(4.1.16) $IR(p/C, r; \alpha) = ru + \theta(ID_o(p/C, u) - \alpha), \alpha > 0.$

71

Since $ID_o(p/C, u)$ is homogeneous of degree $+1$ in outputs, we may write

$$
\begin{aligned}
IR(p/C, r; \alpha) &= ru + \alpha\theta(ID_o(p/C, u/\alpha) - 1) \\
&= \alpha(rv + \theta(ID_o(p/C, v) - 1) \\
&= \alpha IR(p/C, r; 1) = \alpha IR(p/C, r).
\end{aligned}
$$

From above, by substituting $\alpha IR(p/C, r)$ for $IR(p/C, r; \alpha)$ in (4.1.16) and differentiating both sides with respect to α, we find that at an optimum,

(4.1.17) $-\theta = IR(p/C, r).$

Inserting (4.1.17) into (4.1.15) yields

(4.1.18) $IR(p/C, r)ID_o(p/C, u) = ru.$

This expression which is known as the *Cost Indirect Conjugate Duality Theorem*, shows that the product of the indirect revenue and output distance functions equals maximal total revenue. From (4.1.15) we also obtain

(4.1.19) $\nabla_r IR(p/C, r) = u(p/C, r)$

(4.1.20) $\nabla_{p/C} IR(p/C, r) = \theta\nabla_{p/C} ID_o(p/C, u).$

The first expression is the same as expression (4.1.5). The second expression, using (4.1.17) becomes

(4.1.21) $\nabla_{p/C} IR(p/C, r) = IR(p/C, r)\nabla_{p/C} ID_o(p/C, u)$

Combining (4.1.21) and (4.1.8) yields the (compensated) input factor demands

(4.1.22) $x_n(p/C, u) = \varepsilon_c \dfrac{\partial ID_o(p/C, u)}{\partial(p_n/C)}, n = 1, \cdots, N.$

These factor demands differ from those derived from Shephard's lemma in that they are 'compensated' – note the cost-deflated input prices and elasticity term in (4.1.22). From the second part of (4.1.14), we obtain the following Lagrangian problem

(4.1.23) $ID_o(p/C, u) = ru + \theta(IR(p/C, r) - 1).$

72

Applying the same trick as above, we can show that the Lagrangian multiplier θ at optimum equals the distance function, i.e.,

(4.1.24) $\quad \theta = ID_o(p/C, u).$

The derivative properties associated with (4.1.23) are

(4.1.25) $\quad \nabla_u ID_o(p/C, u) = r(p/C, u)$

(4.1.26) $\quad \nabla_{p/C} ID_o(p/C, u) = \theta \nabla_{p/C} IR(p/C, r).$

The first statement shows that the (revenue) deflated cost indirect output shadow price vector equals the gradient with respect to outputs of $ID_o(p/C, u)$. This shadow price vector will be further discussed in Chapter 5. The second expression (4.1.26) can be rewritten, using the expressions (4.1.8) and (4.1.24), as

(4.1.27) $\quad \nabla_{p/C} ID_o(p/C, u) = \dfrac{ID_o(p/C, u) \cdot IR(p/C, r)x}{\varepsilon_c},$

or equivalently as compensated input factor demand equations

(4.1.28) $\quad x_n(p/C, u) = \varepsilon_c \dfrac{\partial ID_o(p/C, u)}{\partial(p_n/C)}, n = 1, \cdots, N,$

where $1 = ID_o(p/C, u)IR(p/C, r)$. To convince oneself that $1 = ID_o(p/C, u) \cdot IR(p/C, r)$, refer to the second expression in (4.1.14).

Finally, one should note that (4.1.18) can be written as

(4.1.29) $\quad IR(p/C, r)ID_o(p/C, u) = \nabla_r IR(p/C, r)\nabla_u ID_o(p/C, u),$

see (4.1.19) and (4.1.25). That is, the product of the indirect revenue and output distance functions equals the inner product of their gradient vectors with respect to output prices and output quantities, respectively.

In Section 3.1, we gave two examples of functional forms which can be used to parameterize the (indirect) output distance function. For completeness we supply two similar examples based on the indirect revenue function. First, the *Translog Indirect Revenue Function* is defined as

(4.1.30) $\quad \ln IR(p/C, r) \ = \ \alpha_0 + \displaystyle\sum_{n=1}^{N} \alpha_n \ln(p_n/C) + \sum_{m=1}^{M} \beta_m \ln r_m$

73

$$+\frac{1}{2}\sum_{n=1}^{N}\sum_{n'=1}^{N}\alpha_{nn'}(\ln(p_n/C))(\ln(p_{n'}/C))$$

$$+\frac{1}{2}\sum_{m=1}^{M}\sum_{m'=1}^{M}\beta_{mm'}(\ln r_m)(\ln r_{m'})$$

$$+\sum_{n=1}^{N}\sum_{m=1}^{M}\gamma_{nm}(\ln(p_n/C))(\ln r_m).$$

The parameters are restricted to ensure that $IR(p/C,r)$ is homogeneous of degree $+1$ in output prices by imposing $\sum_{m=1}^{m}\beta_m = 1, \sum_{m'=1}^{M}\beta_{mm'} = \sum_{m=1}^{M}\gamma_{nm} = 0, m = 1, \cdots, M, n = 1, \cdots, N.$ Symmetry is imposed by $\alpha_{nn'} = \alpha_{n'n}, \beta_{mm'} = \beta_{m'm}, n, n' = 1, \cdots, N, m, m' = 1, \cdots, M.$

Our second parameterization of the indirect revenue function is a *Variation of the Generalized Leontief Indirect Revenue Function*, namely

$$(4.1.31) \qquad IR(p/C,r) = \frac{\sum_{m=1}^{M}\sum_{m'=1}^{M}\beta_{mm'}(r_m r_{m'})^{\frac{1}{2}} + \sum_{m=1}^{M}\sum_{n=1}^{N}\gamma_{nm}(p_n/C)r_m}{\sum_{n=1}^{N}\sum_{n'=1}^{N}\alpha_{nn'}((p_n/C)(p_{n'}/C))^{\frac{1}{2}}}.$$

This function is homogeneous of degree $+1$ in output prices (which the reader should verify), homogeneous of degree zero and nonlinear in the parameters. See Section 3.1 for an example of how this function can be normalized and transformed into a function that is linear in its parameters. Note that (4.1.31) readily admits zero values in the data whereas (4.1.30) does not.

4.2 Cost Minimization

In Section 3.2 the revenue indirect input distance function was introduced as a function representation of the revenue indirect input correspondence. Its dual revenue representation is discussion in this section. In particular, taking revenue deflated output prices as given, we define, for given input prices, the minimum total cost needed to meet a target revenue. This revenue indirect cost function is shown to be dual to the revenue indirect input distance function. Moreover, we derive the properties this cost function inherits from the parent indirect technology, and prove some derivative properties.

Denote output prices by $r = (r_1, \cdots, r_M) \in \Re_+^M$, input prices by $p = (p_1, \cdots, p_N) \in \Re_+^N$, and target revenue by R. We suppose throughout that R is positive. The *Revenue Indirect Cost Function* is defined by

$$(4.2.1) \qquad IC(r/R,p) = \inf_x \{px : x \in IL(r/R)\} = \inf_x \{px : x \in L(u), ru \geqq R\}.$$

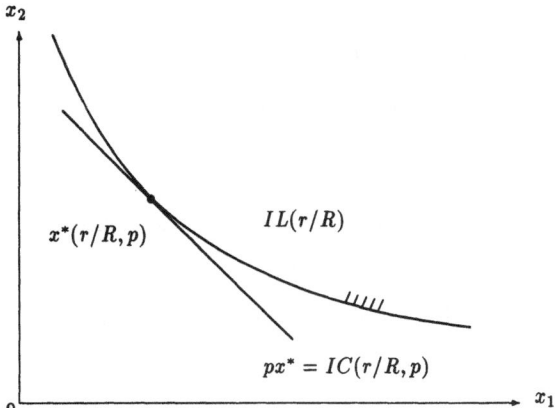

Figure 4.4: The Revenue Indirect Cost Function

We do not assume strict positivity of input prices or output prices, therefore we can not assure
that the infimum is achieved. However, if $p > 0$ and $IL(r/R)$ is nonempty, then the infimum is
achieved and can be replaced by minimum in (4.2.1). To prove that a minimum is achieved under
the above two conditions, let $x^\circ \in IL(r/R)$, since $IL(r/R)$ is closed, IL.5, the intersection
$(IL(r/R) \bigcap \{x : px \leqq px^\circ\})$ is compact $(p > 0)$. Thus the minimum px^* is achieved on this
intersection. Moreover, for any $y \in IL(r/R)$ not in the intersection, $px^* < py$. This proves our
claim.

To illustrate (4.2.1) consider Figure 4.4. Suppose that $p > 0, IL(r/R) \neq \emptyset$ and that two inputs
are used. The revenue indirect input set is denoted by $IL(r/R)$, the price vector is $p = (p_1, p_2)$.
The input vector which minimizes the cost of producing target revenue R at output prices r is
$x^*(r/R, p)$, and the associated minimal cost equals $px^* = IC(r/R, p)$.

Proposition (3.2.13) assures that the indirect input correspondence can be modeled by the
indirect input distance function. Thus we may express the indirect cost function as

(4.2.2) $\qquad IC(r/R, p) = \inf_x \{px : ID_i(r/R, x) \geqq 1\}.$

Expression (4.2.2) shows that the indirect cost function can be derived from the indirect input
distance function. This constitutes one half of the duality between indirect distance and cost

75

functions. The other half shows that the indirect input distance function can be deduced from knowledge about the indirect cost function. That is

(4.2.3) $\quad ID_i(r/R, x) = \inf_p \{px : IC(r/R, p) \geqq 1\}.$

Before we prove the above duality theorem, consider the problem

(4.2.4) $\quad IC(r/R, p) = \min_x \{px : R(x, r) \geqq R\}.$

Proposition (2.4.5) assures that $IL(r/R)$ can be expressed by the revenue function $R(x, r)$ as in (4.2.4), thus if $IL(r/R)$ is nonempty and input prices are positive, (4.2.4) is an accurate expression of indirect cost. The Lagrangian formulation of (4.2.4) is

(4.2.5) $\quad IC(r/R, p) = px + \lambda(1 - R(x, r/R)),$

and the corresponding derivative properties are

(4.2.6) $\quad \nabla_p IC(r/R, p) = x(r/R, p),$

(4.2.7) $\quad \nabla_{r/R} IC(r/R, p) = -\lambda \nabla_{r/R} R(x, r/R).$

The first property yields the vector of revenue indirect factor demand equations, where each demand function $x_n(r/R, p)$ depends on revenue deflated output prices $(r/R) = (r_1/R, \cdots, r_M/R)$ and input prices $p = (p_1, \cdots, p_N)$. The interpretation of (4.2.7) requires some preliminaries. First suppose that Shephard's (output) lemma holds, see Appendix, i.e., we can derive the revenue deflated output supply functions from the revenue function $u(x, r/R) = \nabla_{r/R} R(x, r/R)$. Second, we need to interpret the Lagrangian multiplier λ. This is done by considering the first order conditions associated with (4.2.5). These are

(4.2.8) $\quad$ (a) $\quad p_n = \lambda \dfrac{\partial R(x, r/R)}{\partial x_n}, n = 1, \cdots, N,$

$\quad\quad\quad\quad$ (b) $\quad 1 = R(x, r/R).$

Multiplication of (a) by x_n and summing over $n = 1, \cdots, N$ and using the following definition of scale elasticity $\varepsilon_R(x, r/R) = -\sum_{n=1}^N x_n \frac{\partial R(x, r/R)}{\partial x_n}$, allows us to restate the multiplier as $\lambda = IC/\varepsilon_R$. Note that we assume $R(x, r/R) = 1$ in the definition of elasticity, and that $\sum_{n=1}^N p_n x_n = IC(r/R, p)$, where x is from (4.2.6). Thus we can rewrite (4.2.7) as

(4.2.9) $\quad \nabla_{r/R} IC(r/R, p) = \dfrac{uIC}{\varepsilon_R}$

or equivalently,

$$(4.2.10) \qquad u_m = \frac{\varepsilon_R \nabla_{r_m/R} IC(r/R, p)}{IC}, m = 1, \cdots, M,$$

that is, the revenue indirect supply functions $u_m(r/R, p), m = 1, \cdots, M$, can be derived from the indirect cost function.

Suppose that input prices are positive and that $IL(r/R)$ is nonempty. Let inputs be strongly disposable, i.e, IL.2.S holds, and suppose that properties IL.1 and IL.5 apply, then the indirect cost function is

IC.1   nonnegative and nondecreasing in (positive) input prices,

IC.2   homogeneous of degree +1 in (positive) input prices,

IC.3   concave and continuous in (positive) input prices.

To prove IC.1 note that since $IL(r/R)$ is closed and nonempty,

$IC(r/R, p) = \min_x \{px : x \in IL(r/R)\}$, thus since $IL(r/R) \subseteq \Re_+^N$, and $p > 0$, clearly, $IC(r/R, p) \geqq 0$. Moreover, if $p^o \geqq p$, then cost is nondecreasing. This proves IC.1.

Let $\lambda > 0$, then

$IC(r/R, \lambda p) = \min_x \{\lambda px : x \in IL(r/R)\} = \lambda \min_x \{px : x \in IL(r/R)\} = \lambda IC(r/R, p)$. Thus IC.2 holds.

The proof of IC.3 is deferred to the problem section.

Our next objective is to show that the indirect input distance function can be retrieved from the indirect cost function. First, however, we show that $IL(r/R)$ can be derived from $IC(r/R, p)$, and then the result follows from the fact that $ID_i(r/R, x) = \sup\{\lambda : (x/\lambda) \in IL(r/R)\}$. Thus, define

$$(4.2.11) \qquad IL^*(r/R) = \{x : px \geqq IC(r/R, p), \forall p > 0\} = \bigcap_{p>0} \{x : px \geqq IC(r/R, p)\}.$$

Figure 4.5 illustrates the input set $IL^*(r/R)$. The figure shows two upper halfspaces spanned by the price vectors p^o and p'. Their intersection contains $IL^*(r/R)$. Note that input prices are required to be strictly positive and not semi-positive, as in the case of the output prices in (4.1.10).

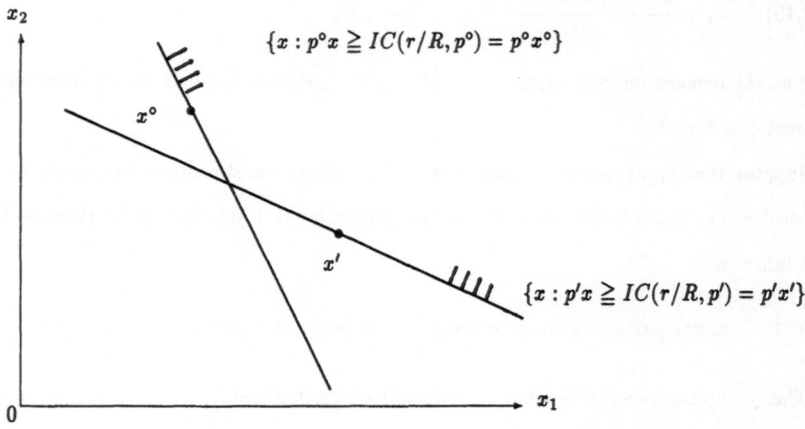

Figure 4.5: Construction of $IL^*(r/R)$

(4.2.12)   **Proposition:** Suppose that the revenue indirect input correspondence satisfies (IL.1, IL.2.S, IL.5) and that it has convex input sets $IL(r/R)$, then

$$\forall (r/R) \in \Re_+^M, IL(r/R) = IL^*(r/R).$$

Proof: If $IL(r/R) = \emptyset$, e.g., if $r = 0$, then $IL(r/R) \subseteq IL^*(r/R)$. Moreover, $IC(r/R,p) = \inf_x\{px : x \in IL(r/R)\} = +\infty$ since $IL(r/R) = \emptyset$, thus $IL^*(r/R) = \emptyset$ and $IL(r/R) = IL^*(r/R)$. Assume that $x^\circ \in IL(r/R)$, then $px^\circ \geq IC(r/R,p), \forall p > 0$. Therefore, $x^\circ \in IL^*(r/R)$. Conversely assume that $x^\circ \notin IL(r/R)$, but $x^\circ \in \Re_+^N$. The last is possible since $r \neq 0$ and by IL.1, $IL(r/R) \neq \Re_+^N$. Since $IL(r/R)$ is closed and convex, by the strict separation theorem (Rockafeller 1970, p. 99), there exists $p^\circ \in \Re^N, p^\circ \neq 0$, and $\varepsilon > 0$ such that $(p^\circ x^\circ + \varepsilon) < \inf_x\{p^\circ x : x \in IL(r/R)\}$. Property IL.2.S implies that $p^\circ \in \Re_+^N$, thus $p^\circ \geq 0$. To prove that p° can be taken to be strictly positive, note that since $\varepsilon > 0$ there exists $p' > 0$ such that $p'x^\circ > \varepsilon/2$. Define $\hat{p} = (p^\circ + p')$, then $\hat{p} > 0$, and $\hat{p}x^\circ < \inf_x\{p^\circ x : x \in IL(r/R)\} \leq \min_x\{\hat{p}x : x \in IL(r/R)\}$, thus $x^\circ \notin IL^*(r/R)$, and $IL(r/R) = IL^*(r/R)$.

$Q.E.D.$

Figure 4.6 illustrates the last proposition. The hyperplane spanned by the price vector p° separates x° from $IL(r/R)$, however $p_2^\circ = 0$, thus a second price vector $\hat{p}$, strictly positive is found

78

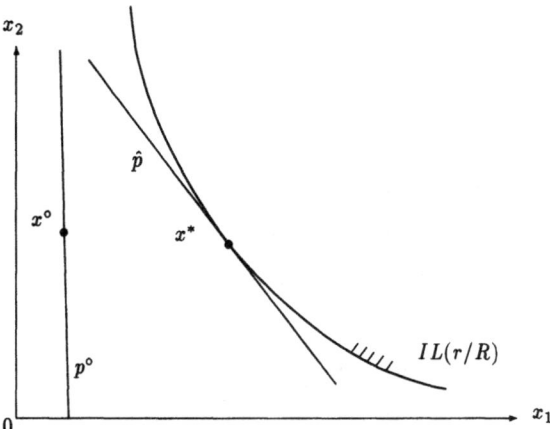

Figure 4.6: Illustration of Proposition (4.2.12)

such that the new hyperplane spanned by it also separates x° from $IL(r/R)$. Note that since $\hat{p} > 0, \min_x\{\hat{p}x : x \in IL(r/R)\}$ is achieved.

Proposition (4.2.12) is the basic tool needed to prove that

$$(4.2.13) \quad ID_i(r/R, x) = \inf_p\{px : IC(r/R, p) \geqq 1\}, x \in \Re_+^N.$$

However, we also make use of

$$(4.2.14) \quad \{x : px \geqq IC(r/R, p), \forall p > 0\} = \{x : px \geqq 1, IC(r/R, p) \geqq 1\}.$$

The proof of (4.2.14) is similar to that of (4.1.13) and is omitted. To prove (4.2.13), let $x \in \Re_+^N$, and $(r/R) \in \Re_+^M$, then

$$
\begin{aligned}
ID_i(r/R, x) &= \sup\{\lambda : (x/\lambda) \in IL(r/R)\} \\
&= \sup\{\lambda : (x/\lambda) \in IL^*(r/R)\} \\
&= \sup\{\lambda : p\frac{x}{\lambda} \geqq IC(r/R, p), \forall p > 0\} \\
&= \sup\{\lambda : px \geqq \lambda, IC(r/R, p) \geqq 1\} \\
&= \inf\{px : IC(r/R, p) \geqq 1\}.
\end{aligned}
$$

Q.E.D.

The two expressions (4.2.3) and (4.2.13) form our duality pair for the revenue indirect distance and cost functions. For reference, it is useful to state these together.

$$(4.2.15) \qquad IC(r/R, p) = \inf_x \{px : ID_i(r/R, x) \geq 1\}$$
$$ID_i(r/R, x) = \inf_p \{px : IC(r/R, p) \geq 1\}.$$

The above duality pair shows that the revenue indirect cost function can be derived from the revenue indirect distance function, and that the converse is also true. If convexity of the indirect input sets $IL(r/R)$ is not imposed, then the second part of (4.2.15) holds with inequality only, i.e., $ID_i(r/R, x) \leq$

$\inf_p\{px : IC(r/R, p) \geq 1\}$. In our proof of duality, we have imposed strong disposability of inputs. However, if we wish to allow for congestion, we would impose weak rather than strong disposability, i.e., IL.2. In that case, we need to allow for some input prices to be negative, otherwise our duality theorem does not hold. Finally, whenever $p > 0$, and $IL(r/R) \neq \emptyset$, we may use minimum rather than infimum, which is also true for the second part when $x > 0$.

To continue the list of properties the indirect cost function inherits from the parent technology, consider

IC.4 $\qquad IC(r/R, 0) = 0$, if $(r/R) \in \Re_{++}^M$.

IC.5 $\qquad IC(0, p) = +\infty, p > 0$.

IC.6 $\qquad$ If $(r/R)^\circ \geq (r/R) > 0$, then $IC(r/R, p) \geq IC((r/R)^\circ, p)$.

The interpretation and proof of the last three properties are left to the reader. Also, the reader should consider the full set of properties which $IC(r/R, p)$ inherits from IL.1-IL.7.

The duality pair (4.2.15) admits the following four derivative properties

$$(4.2.16) \qquad \nabla_p IC(r/R, p) = x(r/R, p)$$

$$(4.2.17) \qquad \nabla_{r/R} IC(r/R, p) = -\lambda \nabla_{r/R} ID_i(r/R, x)$$

$$(4.2.18) \qquad \nabla_x ID_i(r/R, x) = p(r/R, x)$$

$$(4.2.19) \qquad \nabla_{r/R} ID_i(r/R, x) = - \wedge \nabla_{r/R} IC(r/R, p)$$

80

where λ is the Lagrangian to the following problem

(4.2.20) $IC(r/R, p) = px + \lambda(1 - ID_i(r/R, x))$

and $\wedge$ is the Lagrangian for

(4.2.21) $ID_i(r/R, x) = px + \wedge(1 - IC(r/R, p)).$

The first derivative expression (4.2.16) yields the factor demands associated with indirect cost minimization. Following the method developed in Section 4.1, one can show that $\lambda = IC(r/R, p)$, thus the second expression may be written without invoking the Lagrangian. In a similar way, expression (4.2.19) can be written without its Lagrangian $\wedge$. The third expression yields the cost deflated input prices obtained as the gradient with respect to inputs. This may be termed the cost indirect dual Shephard's lemma. Similarly, (4.2.16) would be the cost indirect Shephard's lemma.

Two parametric examples of the revenue indirect cost function conclude this section. First the translog formulation is

$$
(4.2.22) \quad \ln IC(r/R, p) = \alpha_0 + \sum_{m=1}^{M} \alpha_m \ln(r_m/R) + \sum_{n=1}^{N} \beta_n \ln p_n
$$
$$
+ \frac{1}{2} \sum_{m=1}^{M} \sum_{m'=1}^{M} \alpha_{mm'}(\ln(r_m/R))(\ln(r_{m'}/R))
$$
$$
+ \frac{1}{2} \sum_{n=1}^{N} \sum_{n'=1}^{N} \beta_{nn'}(\ln p_n)(\ln p_{n'})
$$
$$
+ \sum_{m=1}^{M} \sum_{n=1}^{N} \gamma_{mn}(\ln(r_m/R))(\ln p_n),
$$

with the usual parameter restrictions, which we leave to the reader to specify. The second parametric form is our variation of the Generalized Leontief functional form, namely

$$
(4.2.23) \quad IC(r/R, p) = \frac{\sum_{n=1}^{N} \sum_{n'=1}^{N} \beta_{nn'}(p_n p_{n'})^{\frac{1}{2}} + \sum_{m=1}^{M} \sum_{n=1}^{N} \gamma_{mn}(r_m/R)p_n}{\sum_{m=1}^{M} \sum_{m'=1}^{M} \alpha_{mm'}((r_m/R)(r_{m'}/R))^{\frac{1}{2}}}
$$

This function is homogeneous of degree one in input prices without further restrictions, and homogeneous of degree zero in its parameters. See Section 3.1 for how it may be normalized and made into a function that is linear in its parameters. Note that (4.7.23) allows for inclusion of data with zeros, whereas the translog form does not.

81

4.3 Profit Maximization

The two preceding sections analyzed expenditure and revenue constrained optimization. In this section, we use those models to discuss profit maximization. A simple example illustrates. Suppose the technology can be expressed by the production function

(4.3.1) $u = x^{\frac{1}{2}}$.

The corresponding cost function is

(4.3.2) $C(u,p) = pu^2$,

and the cost indirect revenue function is given by

$$(4.3.3) \quad \begin{aligned} IR(p/C,r) &= \max\{ru : pu^2 \leqq C\} \\ &= \max\{ru : u \leqq 1/(p/C)^{\frac{1}{2}}\} \\ &= r/(p/C)^{\frac{1}{2}}. \end{aligned}$$

Now consider the maximization problem

(4.3.4) $\max\limits_{C} IR(p/C,r) - C$.

Problem (4.3.4) expresses the maximum difference between indirect revenue and target cost. In our example, the maximum equals $(r^2/4p)$, and it is achieved at cost $C = (r^2/4p)$ and indirect revenue $IR(p/C,r) = (r^2/2p)$. The question now arises, is this maximum a usual profit maximum. To answer this question, consider the profit maximization problem

$$(4.3.5) \quad \begin{aligned} \Pi(r,p) &= \max\limits_{(x,u)}\{ru - px : D_o(x,u) \leqq 1\} \\ &= \max\limits_{(x,u)}\{rx^{\frac{1}{2}} - px\}. \end{aligned}$$

The maximum for our particular example equals $\Pi(r,p) = (r^2/4p)$, and it is achieved at cost $px = (r^2/4p)$ and (direct) revenue $ru = (r^2/2p)$. Thus for this particular example, the two problems (4.3.4) and (4.3.5) have the same solution. In general, when a maximum exists, we can prove the following

(4.3.6)   **Proposition:** Suppose that (4.3.4) and (4.3.5) have solutions, then their solutions are equal.

Proof: Assume that $\max_{(x,u)}\{ru - px : D_o(x,u) \leq 1\} > \max_C\{IR(p/C,r) - C\}$, i.e., $ru^* - px^* > IR(p/C^*,r) - C^*$, where "*" denotes the optimizer, and where $IR(p/C^*,r) = r\hat{u} - C^* = r\hat{u} - p\hat{x}$, where $\hat{u}$ is the optimizer for $IR(p/C^*,r)$ and $C^* = p\hat{x}$. Thus we have that $ru^* - px^* > r\hat{u} - p\hat{x}$. Now, $D_o(x^*,u^*) \leq 1$, and $px^* \leq p^*x = C$, thus since $r\hat{u} - p\hat{x} \geq ru - px$ for all (x,u) such that $D_o(x,u) \leq 1$, and $px \leq C$ for some $C \geq 0$, $r\hat{u} - p\hat{x} \geq ru^* - px^*$. This leads to the contradiction $ru^* - px^* > ru^* - px^*$, and proves that

$\max_{(x,u)}\{ru - px : D_o(x,u) \leq 1\} \leq \max_C\{IR(p/C,r) - C\}$. To prove the converse inequality, suppose that $\max_C\{IR(p/C,r) - C\} > \max_{(x,u)}\{ru - px : D_o(x,u) \leq 1\}$, i.e.,

$r\hat{u} - \hat{C} = r\hat{u} - p\hat{x} > ru^* - px^*$. Then, since $ru^* - px^* \geq ru - px$ for all (x,u) such that $D_o(x,u) \leq 1$, $ru^* - px^* \geq r\hat{u} - p\hat{x}$, where $p\hat{x} = \hat{C} \geq 0$. This leads to the contradiction $r\hat{u} - p\hat{x} > r\hat{u} - p\hat{x}$, and ends the proof.

$$Q.E.D.$$

Proposition (4.3.6) shows that whenever the two maximization problems (4.3.4) and (4.3.5) have solutions, their two maxima are equal. A difference between the usual profit maximization and evaluation of

(4.3.7) $\qquad IR(p/C,r) - C,$

is highlighted by our next example, in which we suppose that the technology can be modeled by $u = x^2$. The profit maximization problem then becomes $\sup_x\{rx^2 - px\}$. For simplicity suppose that $p = r = 1$, then $\sup_x\{x^2 - x\} = +\infty$. On the other hand, $IR(p/C,r) = r/(p/C)^2$, and for $p = r = 1$, (4.3.7) equals

(4.3.8) $\qquad r/(p/C)^2 - C = C^2 - C$

Thus the difference $IR(p/C,r) - C$ is defined even for the technology $u = x^2$ (which exhibits everywhere increasing returns to scale). On the other hand, $\sup_C IR(p/C,r) - C$ is, of course, unbounded.

Let us consider a third profit maximization problem

(4.3.9) $\qquad \Pi(r,p/C) = \max_{(x,u)}\{ru - px : px \leq C, D_o(x,u) \leq 1\}.$

In this problem, assuming that a maximum is achieved, profit maximization $(ru - px)$ is restricted by both the technology $D_o(x,u) \leq 1$, and expenditure or budget $px \leq C$. The problem has been termed *Cost (Expenditure) Constrained Profit Maximization*, and it is applicable to firms with restricted budgets. Sometimes, the budget constraint only applies to a subvector of inputs, say x_S. The problem then becomes a short run expenditure constrained profit maximization problem, i.e.,

$$(4.3.10) \quad \Pi(r, p_S/C, x_{\hat{S}}) = \max_{(x_S, u)} \{ru - p_S x_S : p_S x_S \leqq C, D_o(x_S, x_{\hat{S}}, u) \leqq 1\},$$

Where $x = (x_S, x_{\hat{S}})$, and $x_{\hat{S}}$ may be fixed inputs, e.g.

If profit $\Pi(r, p/C)$ is maximized with respect to cost C, i.e.,

$$(4.3.11) \quad \max_C \Pi(r, p/C),$$

and if the maximum exists, $\Pi(r, p) = \max_C \Pi(r, p/C)$, i.e., the maximum of (4.3.10) equals the usual profit maximum. The proof to the last claim is left to the reader; here we turn to profit maximization given the revenue indirect cost function. A simple example illustrates.

Suppose again that the technology is modeled by the production function $u = x^{\frac{1}{2}}$. The corresponding revenue function takes the form

$$(4.3.12) \quad R(x, r) = r x^{\frac{1}{2}},$$

and the revenue indirect cost function equals

$$(4.3.13) \quad IC(r/R, p) = p/(r/R)^2.$$

Now consider the maximization problem

$$(4.3.14) \quad \max_R R - IC(r/R, p).$$

This problem expresses the maximal difference between target revenue R and revenue indirect cost $IC(r/R, p)$. In our simple example, the maximum over R exists and equals $(r^2/4p)$. The maximum is attained at revenue $R = (r^2/2p)$ and indirect cost $IC(r/R, p) = (r^2/4p)$. We observe that the outcome of (4.3.13) with $u = x^{\frac{1}{2}}$ coincides with the maximization problem (4.3.4) and (4.3.5). In general one can prove

$(4.3.15)$ **Proposition:** Suppose that (4.3.5) and (4.3.4) have solutions, then their solutions are equal.

84

The proof of Proposition (4.3.5) is left to the reader. Thus far we have shown that whenever a profit maximum exists, it can be obtained from the cost indirect problem, the revenue indirect problem and the cost constrained maximization problem (4.3.8). The last observation hints toward the formulation of the revenue constrained maximization problem

$$(4.3.16) \quad \Pi(r/R, p) = \max_{(x,u)} \{ru - px : ru \geq R, D_o(x, u) \leq 1\}.$$

In the last problem, assuming that a maximum is achieved, profit maximization $(ru - px)$ is restricted by revenue $ru \geq R$, and the technology $D_o(x, u) \leq 1$. Clearly, one may assume that only a subvector u_S is revenue restricted, i.e., $r_S u_S \geq R$. As in the case of the profit maximization problem (4.3.8), whenever $\max_R \Pi(r/R, p)$ exists, this maximum equals the usual profit maximization problem (4.3.5).

Appendix: Cost and Revenue Dualities

The purpose of this appendix is to formulate the duality theorems between cost and input distance functions and between revenue and output distance functions. Also the corresponding Shephard's lemmata are stated.

The cost function can be defined in terms of the input distance function, since $x \in L(u) \iff D_i(u, x) \geq 1$. Thus when the distance function can be retrieved from the cost function we have the first of the two duality theorems of this Appendix

$$(4.A.1) \quad C(u, p) = \inf_x \{px : D_i(u, x) \geq 1\}$$
$$D_i(u, x) = \inf_p \{px : C(u, p) \geq 1\}.$$

The proof of (4.A.1) which requires that the input sets $L(u), u \in \mathcal{R}_+^M$ are convex, is similar to the proof of the duality (4.2.15) and is left to the reader.

From (4.A.1) the following two Shephard's lemmata can be derived.

$$(4.A.2) \quad \nabla_p C(u, p) = x(u, p)$$

$$(4.A.3) \quad \nabla_x D_i(u, x) = p(u, x).$$

(4.A.2) is the traditional Shephard's lemma, and it states that the factor demands $x_n(u, p), n = 1, \cdots, N$, can be derived as the derivative of the cost function $\frac{\partial C(u, p)}{\partial p_n}, n = 1, \cdots, N$.

The second set of derivative properties, the (input price) Shephard's lemma states that cost deflated shadow input prices $p_n(u, x), n = 1, \cdots, N$, can be computed as a derivative of the input distance function. It is important to note that these virtual or support prices are cost deflated. The second duality pair is between revenue and output distance functions, in particular

(4.A.4) $\quad R(x, r) = \sup_u \{ru : D_o(x, u) \leqq 1\}$
$\qquad D_o(x, u) = \sup_r \{ru : R(x, r) \leqq 1\}.$

This duality pair states that the revenue function can be computed from the output distance function, and that the output distance function can be retrieved from the revenue function by "maximization" over output prices. The two corresponding Shephard's lemmata are

(4.A.5) $\quad \nabla_r R(x, r) = u(x, r)$
$\qquad \nabla_u D_o(x, u) = r(x, r)$

The top of (4.A.5) states that the output supply functions $u_m(x, r), m = 1, \cdots, M$, can be obtained from the revenue function as its derivative with respect to the corresponding output prices. The bottom of (4.A.5) shows that revenue deflated output shadow prices are the derivatives of the output distance function with respect to the corresponding output quantities. Note that these virtual or support prices are revenue deflated.

4.P Problems

(4.P.1) The two output supply vectors (4.1.5) (derived from the indirect revenue function) and (4.A.5) both depend on output prices. However (4.1.5) depends also on cost deflated input prices while (4.A.5) depends on inputs. Discuss under what conditions the supply vectors are equal.

(4.P.2) Suppose that the output set is not convex. Show by example that Proposition (4.1.11) does not hold.

(4.P.3) Prove that the inequality $ID_o(p/C, u) \geqq \max_r \{ru : IR(p/C, r) \leqq 1\}$ holds without convexity of $IP(p/C)$.

(4.P.4) Prove the two properties IR.1 and IR.2.

(4.P.5)   Show that the two indirect revenue functions (4.1.30) and (4.1.31) are homogeneous in output prices.

(4.P.6)   In the definition of indirect cost (4.2.1), show that if the input correspondence $L(u)$ is piecewise linear, see Chapter 2, then a minimum can be used in place of infimum.

(4.P.7)   Construct an example showing that Proposition (4.2.12) does not hold if $IL(r/R)$ is not convex.

(4.P.8)   Find conditions on $P(x)$ gauranteeing that $IL(r/R)$ is convex.

(4.P.9)   Prove the claim (4.2.14).

(4.P.10)   Prove properties IC.3 – IC.5.

(4.P.11)   Verify that $\nabla_x ID_i(r/R, x) = p$, and find the cost by which p is normalized.

(4.P.12)   Discuss the parameter restrictions that must be imposed on (4.2.23).

(4.P.13)   Prove that $\Pi(r, p) = \max_C \Pi(r, p/C)$.

(4.P.14)   Prove Proposition (4.3.15).

(4.P.15)   Sketch a proof of the duality pair (4.A.1).

(4.P.16)   Verify that the vector $r(x, r)$ in (4.A.5) are revenue deflated output prices.

Notes

The material in this chapter extends Shephard (1970; 1974), and Shephard and Färe (1980).

87

Chapter 5

Applications of Duality

5.0   Introduction

Duality theory has proved extremely useful both as an analytical device and as an aid to the applied researcher. With respect to the latter, duality accommmodates a wide range of behavioral objectives, including the budget and revenue constrained cases which are the focus of this study. By developing equivalent representations of technology under varying objectives, duality affords the researcher a range of options which differ in their data requirements. Finally, duality can be exploited to derive shadow values or virtual prices. The purpose of this chapter is to elaborate on these points.

We begin by developing the dualities between direct and indirect representations of technology. Included are dualities between the following pairs: (1) cost indirect and direct output distance functions, and (2) revenue indirect and direct input distance functions. We also establish a (one way) relation between the following pairs: (1) cost indirect and direct revenue functions, and (2) revenue indirect and direct cost functions.

One reason for establishing dual relationships between various function representations is to be able to retrieve information which cannot be "obtained" directly from available data. The most well known example is the application of Shephard's lemma to the cost function to derive input demands. Shephard's lemma allows one to retrieve information on the optimal quantities which "support" observed relative prices. In Section 5.2 we apply Shephard type lemmas to retrieve the shadow or virtual prices which support observed output and input choices. This allows us to derive shadow prices of goods (or bads) which may be nonmarketable such as public services or externalities. One may also solve for "undistorted" shadow prices which reflect resource use for

88

inputs or outputs which have distorted observed prices due to regulation or lack of competition, for example.

Section 5.2 shows how to derive shadow prices for outputs and inputs from some of our indirect models. Three methods of deriving shadow prices of outputs are discussed based on: (1) the indirect revenue function and indirect output distance function, (2) the direct and indirect input distance functions and (3) the direct and indirect cost functions. A translog specification of the indirect cost function is included which could be used to calculate output shadow prices. Three methods of deriving shadow price of inputs are discussed based on: (1) the indirect cost and indirect input distance functions, (2) the direct and indirect output distance functions, and (3) the direct and indirect revenue functions. Included is a parameterization of the indirect revenue function as translog.

The final section of this chapter focuses on the notions of scale elasticity and returns to scale. Scale elasticity gives a local measure of the scale properties of the technology. In this manuscript we have introduced a variety of representations of technology, accordingly, we develop a variety of measures of scale elasticity, and establish that they are equivalent. In addition this section discusses the global notion of returns to scale with respect to the output set $P(x)$ and the output distance function.

A set of problems and notes concludes.

5.1   Dualities Between Direct and Indirect Production

Of the four duality results discussed in Chapter 4, the first establishes the link between the indirect revenue and indirect output distance functions, the second establishes the link between indirect cost and indirect input distance functions. The last two pairs (discussed in the Appendix) establish links between the revenue function, cost function and the corresponding (direct) distance functions. In this section we will prove duality theorems between the direct and indirect functions, including those between the direct and indirect distance functions. These duality theorems link the direct and indirect representations of technology. Thus, together with the inverse relationship between the (direct) input and output distance function, we have established that each of the eight representations of technology can be retrieved from any of the others.

Suppose that cost deflated input prices (p/C) are positive and that the ray $\lambda u, \lambda > 0$, through the output vector $u \geq 0$ intersects the cost indirect output set $IP(p/C)$. The indirect output distance function may then be defined as

(5.1.1) $\qquad ID_o(p/C, u) = \min\{\theta : (u/\theta) \in IP(p/C)\}.$

To establish the relation between the indirect output distance function $ID_o(p/C, u)$, and the (direct) output function $D_o(x, u)$, we make use of the definition
$IP(p/C) = \{u : D_o(x, u) \leq 1, px \leq C\}$, then,

$$
\begin{aligned}
ID_o(p/C, u) &= \min_{\theta, x}\{\theta : D_o(x, u/\theta) \leq 1, px \leq C\} \\
&= \min_{\theta, x}\{\theta : D_o(x, u) \leq \theta, px \leq C\} \\
&= \min_{x}\{D_o(x, u) : px \leq C\}
\end{aligned}
$$

or

(5.1.2) $\qquad ID_o(p/C, u) = \min_{x}\{D_o(x, u) : px \leq C\}.$

Expression (5.1.2) shows that the indirect output distance function is derived from the (direct) output distance function as the minimum of input vectors costing no more than C. In general the following inequality between the two distance functions holds,

(5.1.3) $\qquad D_o(x, u) \geq ID_o(p/C, u)$, for $px \leq C$.

The last expression is established from the definition of $D_o(x, u)$ and the observation that $P(x) \subseteq IP(p/C)$ for all input vectors such that $px \leq C$. In particular,

$$
\begin{aligned}
D_o(x, u) &= \inf_{\theta}\{\theta : (u/\theta) \in P(x)\} \\
&\geq \inf_{(\theta, x)}\{\theta : (u/\theta) \in P(x), px \leq C\} \\
&= \inf_{\theta}\{\theta : (u/\theta) \in IP(p/C)\} \\
&= ID_o(p/C, u).
\end{aligned}
$$

Q.E.D.

The inequality (5.1.3) and expression (5.1.2) suggest that the direct output distance function is the maximum over cost deflated input prices of the corresponding indirect distance function.

(5.1.4) $\qquad D_o(x, u) = \max_{(p/C)}\{ID_o(p/C, u) : px \leq C\}.$

90

The pair of expressions (5.1.2) and (5.1.4) establish the duality between output distance functions, i.e.,

$$(5.1.5) \qquad ID_o(p/C, u) = \min_x \{D_o(x, u) : px \leq C\}$$

$$D_o(x, u) = \max_{p/C} \{ID_o(p/C, u) : px \leq C\}.$$

To prove the duality statement (5.1.5) we need only show that the direct output distance function can be recovered from the indirect distance function. Here we restrict our proof to the case where the technology exhibits constant returns to scale (see Notes for a more general proof) and where maximum is achieved. Thus consider

$$
\begin{aligned}
(5.1.6) \quad \max_{p/C}\{ID_o(p/C, u) : px \leq C\} &= \max_{p/C}\{C(u, p/C) : px \leq C\} \quad \text{by } (3.1.12) \\
&= \max_{p/C}\{1/C : C(u, p/C) \geq 1/C, px \leq C\} \\
&= \max_{p/C}\{1/C : C(u, p) \geq 1, px \leq C\} \quad \text{by } (C.2) \\
&= \max_{p}\{1/px : C(u, p) \geq 1\} \\
&= \left(\min_{p}\{px : C(u, p) \geq 1\}\right)^{-1} \\
&= 1/D_i(u, x) \quad \text{by } (4.A.1) \\
&= D_o(x, u) \quad \text{by } (3.2.5)
\end{aligned}
$$

$$Q.E.D.$$

The second duality pair of this section is between the direct and indirect input distance functions. Recall that $IL(r/R) = \{x : D_i(u, x) \geq 1, ru \geq R\}$, then by the definition of the revenue indirect input distance function it follows that

$$
\begin{aligned}
ID_i(r/R, x) &= \max_{\lambda, u}\{\lambda : D_i(u, x/\lambda) \geq 1, ru \geq R\} \\
&= \max_{\lambda, u}\{\lambda : D_i(u, x) \geq \lambda, ru \geq R\} \\
&= \max_{u}\{D_i(u, x) : ru \geq R\},
\end{aligned}
$$

i.e.,

$$(5.1.7) \qquad ID_i(r/R, x) = \max_{u}\{D_i(u, x) : ru \geq R\}.$$

Expression (5.1.7) proves that the indirect input distance function can be derived from the direct distance function through maximization with respect to output, with outputs being revenue restricted.

In general, it is true that

(5.1.8) $\quad ID_i(r/R, x) \geqq D_i(u, x)$, for $ru \geqq R$.

This claim is verified by

$$
\begin{aligned}
D_i(u, x) &= \sup\{\lambda : (x/\lambda) \in L(u)\} \\
&\leqq \sup\{\lambda : (x/\lambda) \in L(u), ru \geqq R\} \\
&= ID_i(r/R, x),
\end{aligned}
$$

where the inequality holds since $L(u) \subseteqq IL(r/R)$ for $ru \geqq R$.

The inequality (5.1.8) together with (5.1.7) justify the duality pair

(5.1.9) $\quad ID_i(r/R, x) = \max_u \{D_i(u, x) : ru \geqq R\}$

$\qquad\qquad\qquad D_i(u, x) = \min_{r/R} \{ID_i(r/R, x) : ru \geqq R\}.$

The duality pair (5.1.9) states that the indirect output distance function can be computed from the direct distance function as the maximum of $D_i(u, x)$ over revenue restricted outputs, and that the direct output correspondence can be recovered from the indirect by minimizing it over revenue constrained deflated output prices, i.e., $(r/R)u \geqq 1$. The first part has been proved, the second is left to the reader. We note however that if the output set $P(x)$ is not convex the last part of (5.1.9) holds only with inequality, i.e.,

$$
D_i(u, x) \leqq \min_{r/R} \{ID_i(r/R, x) : ru \geqq R\}.
$$

We now turn to the relation between the direct and the cost indirect revenue function, where the cost indirect revenue function is defined by

$$
IR(p/C, r) = \max_u \{ru : u \in IP(p/C)\},
$$

see (4.1.1). This function may also be written as

(5.1.10) $\quad IR(p/C, r) = \max_{x,u} \{ru : D_o(x, u) \leqq 1, px \leqq C\}$

$\qquad\qquad\qquad\quad = \max_{x,u} \{ru : D_o(x, (ru)u) \leqq ru, px \leqq C\}$

$\qquad\qquad\qquad\quad = \max_{x,u} \{ru : D_o(x, (ru)u) \leqq ru \leqq R(x, r), px \leqq C\}$

$\qquad\qquad\qquad\quad = \max_x \{R(x, r) : px \leqq C\},$

i.e.,

$$(5.1.11) \quad IR(p/C, r) = \max_x \{R(x, r) : px \leqq C\}.$$

The second equality holds due to homogeneity in the output distance function, and since $ru \leqq R(x, r)$ for all $r \in \Re_+^M$, the third equality in (5.1.10) holds.

Expression (5.1.11) shows that the indirect revenue function can be derived from maximization of the direct revenue function over cost constrained inputs. In general, for input vector x such that $px \leqq C$,

$$(5.1.12) \quad IR(p/C, r) \geqq R(x, r).$$

This follows directly from the fact that $P(x) \subseteq IP(p/C)$, whenever $px \leqq C$.

The last relation studied in this section is between the direct and indirect cost functions. Since $L(u) \subseteq IL(r/R)$, for $ru \geq R$, it follows directly that

$$(5.1.13) \quad C(u, p) \geqq IC(r/R, p), \text{ whenever } ru \geq R.$$

Moreover, from the above discussion, it is clear that

$$(5.1.14) \quad IC(r/R, p) = \min_u \{C(u, p) : ru \geq R\}.$$

We end this section with a simple example. Let the production technology be described by $u = x^2$, that is, the technology is homogeneous of degree $+2$. The corresponding cost function $C(u, p)$ equals $p\sqrt{u}$. Now let us compute the indirect cost function according to (5.1.14), then

$$IC(r/R, p) = \min_u \{p\sqrt{u} : ru \geq R\} = p\sqrt{(1/r/R)}.$$

Next given the indirect cost function, the direct cost function may be recovered from

$$C(u, p) = \max_{r/R} \{p\sqrt{(1/r/R)} : ru \geq R\} = p\sqrt{u}.$$

Clearly, the above example is very simple, but it shows the thrust of the relationship between the direct and indirect cost function without imposing constant returns to scale.

93

5.2 Shadow Pricing Inputs and Outputs

We next exploit the relationships derived in the previous section in order to derive shadow or virtual prices which support the observed output and input choices. Our shadow pricing techniques are the mirror image of Shephard's lemma which derives optimal input demands based on observed input prices via the cost function. Here we take advantage of the dual relationship between various distance functions and direct and indirect value functions.

Our objective is to derive undeflated shadow prices that reflect resource use, and we start by shadow pricing outputs, making use of the duality pair (4.1.14)

$$(5.2.1) \qquad \begin{aligned} IR(p/C, r) &= \max_u \{ru : ID_o(p/C, u) \leqq 1\} \\ ID_o(p/C, u) &= \max_r \{ru : IR(p/C, r) \leqq 1\}. \end{aligned}$$

Although we proved this duality under the condition of strong disposability of outputs, here we only require weak disposability. This weaker assumption allows us in particular to shadow price "bads" whenever such outputs exist. Also, as discussed in Section 4.1, the duality theorem still applies given weak output disposability.

Suppose that the cost indirect output distance function is differentiable, and consider the Lagrangian expression associated with the first part of (5.2.1), i.e.,

$$(5.2.2) \qquad \wedge = ru + \theta(ID_o(p/C, u) - 1).$$

The first order conditions with respect to outputs are

$$(5.2.3) \qquad r = -\theta \nabla_u ID_o(p/C, u).$$

It was shown, see (4.1.17), that the Lagrangian multiplier θ equals $-IR(p/C, r)$, thus (5.2.3) takes the form

$$(5.2.4) \qquad r = IR(p/C, r)\nabla_u ID_o(p/C, u).$$

In order to interpret (5.2.4), we use the second part of (5.2.1) and observe that

$$(5.2.5) \qquad \nabla_u ID_o(p/C, u) = r^*(p/C, u),$$

where $r^*(p/C, u)$ is the solution price vector, see also (4.1.24). Thus,

$$(5.2.6) \qquad r = IR(p/C, r)r^*(p/C, u),$$

gives us an expression for output prices r. However, the indirect revenue function $IR(p/C, r)$ is homogeneous of degree $+1$ in the output prices we seek, see IR.2, thus without independent information on $IR(p/C, r)$, we can only identify relative (or indirect revenue deflated) output prices. Given that information on cost deflated input prices and output quantities are available, the indirect distance function can be computed, thus the revenue deflated output price vector $r^*(p/C, u)$ can be derived. To obtain undeflated shadow prices, additional information regarding indirect revenue is required. One assumption that suffices is

(5.2.7)   **Assumption**: The firm is not-for-profit.

This assumption is interpreted as assuming that costs C, which are known, equal indirect revenue $IR(p/C, u)$. Given that Assumption (5.2.7) is applicable, then each output price r_m can be derived according to

$$(5.2.8) \qquad r_m = Cr_m^*(p/C, u), m = 1, \cdots, M,$$

where we have substituted C for $IR(p/C, r)$ in (5.2.6), allowing us to solve for 'undeflated' or absolute shadow prices.

   An alternative way to retrieve undeflated output prices from expression (5.2.6) is to invoke

(5.2.9)   **Assumption**: For $M > 2$, one output shadow price equals its market (observed) price.

This assumption supposes that one shadow output price r_m equals market price say r_m^m. If this assumption applies to the case in question, then $IR(p/C, u)$ can be computed by substituting r_m^m in (5.2.6) and rearranging to yield

$$(5.2.10) \qquad IR^* = r_m^m / r^*(p/C, u).$$

Since IR^* is the same for all outputs $m = 1, \cdots, M$, we may substitute IR^* into (5.2.6) to calculate the undeflated shadow output prices

$$(5.2.11) \qquad r_{m'} = IR^* r_{m'}^*(p/C, u), m' = 1, \cdots, M, m' \neq m.$$

The second approach to shadow pricing outputs invokes the relation between direct and indirect input distance functions. The Lagrangian problem associated with $ID_i(r/R, x)$ in (5.1.9) is

$$(5.2.12) \qquad \wedge = D_i(u, x) + \theta\left(\frac{ru}{R} - 1\right).$$

95

The first order conditions are

$$\nabla_u D_i(u, x) = -\theta r/R$$

$$ru = R.$$

From the first expression, after multiplying both sides by u_m and $1/D_i(u, x)$, then summing over m, we obtain

$$\sum_{m=1}^{M} \frac{u_m \partial D_i(u, x)}{D_i(u, x) \partial u_m} = -\frac{\theta}{D_i(u, x)} \sum_{m=1}^{M} \frac{r_m u_m}{R}.$$

The left hand side is an expression of scale elasticity $\varepsilon_{D_i} = \sum_{m=1}^{M} \frac{u_m \partial D_i(u, x)}{D_i(u, x) \partial u_m}$ (see 5.3.5), and on the right hand side $\sum_{m=1}^{M} r_m u_m = ru = R$, thus, the Lagrangian θ may be expressed as

(5.2.13) $-\theta = D_i(u, x)\varepsilon_{D_i}.$

Using the first order conditions $\nabla_u D_i(u, x)$ together with (5.2.13) yields

(5.2.14) $r_m = \dfrac{R}{\varepsilon_{D_i}} \dfrac{\partial D_i(u, x)}{D_i(u, x)\partial u_m}, m = 1, \cdots, M.$

If information on (x, u) is available, then $D_i(u, x)$ can be estimated, thus one can obtain (r_m/R) for each $m = 1, \cdots, M$. To see this, rearrange (5.2.14). If in addition Assumption (5.2.9) applies or if R or a proxy is known, then the absolute output shadow prices r_m can be derived using (5.2.14).

Our third method for shadow pricing outputs makes use of the relation between the direct cost function and the indirect cost function, see the formulation of $IC(r/R, p)$ in (5.1.14). It may be written in Lagrangian form as

(5.2.15) $\wedge = C(u, p) + \theta(\dfrac{ru}{R} - 1).$

The first order conditions are

$$\nabla_u C(u, p) = -\theta r/R$$

$$ru = R.$$

From the first part, after multiplying by u_m and dividing by $C(u, p)$, we obtain

$$\frac{u_m \partial C(u, p)}{C(u, p)\partial u_m} = -\frac{\theta}{C(u, p)}\frac{r_m u_m}{R}, m = 1, \cdots, M.$$

96

Summing over m and observing that $\varepsilon_C = \sum_{m=1}^{M} \frac{u_m \partial C(u,p)}{C(u,p)\partial u_m}$ (see 5.3.7) the Lagrangian θ may be expressed as

$$(5.2.16) \quad -\theta = C(u,p)\varepsilon_c.$$

This expression together with the first order conditions for (5.2.15) yields

$$(5.2.17) \quad r_m = \frac{R}{\varepsilon_C C(u,p)} \frac{\partial C(u,p)}{\partial u_m}, m = 1, \cdots, M.$$

As with our other two approaches to shadow pricing outputs, one may either calculate deflated shadow prices $r_m/R, m = 1, \cdots, M$ or use one of our assumptions to determine R and solve for undeflated shadow prices.

Next we turn to a more concrete example of how one may calculate output shadow prices. We return to our first approach which was based on the relation between $ID_o(p/C, u)$ and $IR(p/C, r)$. As an example of how to calculate shadow prices for outputs, we suppose that the cost indirect output distance function $ID_o(p/C, u)$ can be parameterized as a translog function. Moreover, we assume that there exist $k = 1, \cdots, K$ observations of positive cost deflated input prices $(p/C)^k = (p_{k1}/C_k, \cdots, p_{kM}/C_k)$ and positive output quantities $u^k = (u_{k1}, \cdots, u_{kM})$. The parameters of the translog form of $ID_o(p/C, u)$ may be computed as the solution to the following Aigner-Chu type linear programming problem

$$
\begin{aligned}
(5.2.18) \quad \max_{\alpha, \beta, \gamma} \quad \sum_{k=1}^{K} & \Bigg[\alpha_0 + \sum_{n=1}^{N} \alpha_n \ln(p_{kn}/C_k) + \sum_{m=1}^{M} \beta_m \ln u_{km} \\
& + \frac{1}{2} \sum_{n=1}^{N} \sum_{n'=1}^{N} \alpha_{nn'} (\ln(p_{kn}/C_k))(\ln(p_{kn'}/C_k)) \\
& + \frac{1}{2} \sum_{m=1}^{M} \sum_{m'=1}^{M} \beta_{mm'} (\ln u_{km})(\ln u_{km'}) \\
& + \sum_{n=1}^{N} \sum_{m=1}^{M} \gamma_{nm} (\ln(p_{kn}/C_k))(\ln u_{km}) \Bigg]
\end{aligned}
$$

$$
\begin{aligned}
\text{subject to} \quad (i) \quad & \alpha_0 + \sum_{n=1}^{N} \alpha_n \ln(p_{kn}/C_k) + \sum_{m=1}^{M} \beta_m \ln u_{km} \\
& + \frac{1}{2} \sum_{n=1}^{N} \sum_{n'=1}^{N} \alpha_{nn'} (\ln(p_{kn}/C_k))(\ln(p_{kn'}/C_k)) \\
& + \frac{1}{2} \sum_{m=1}^{M} \sum_{m'=1}^{M} \beta_{mm'} (\ln u_{km})(\ln u_{km'})
\end{aligned}
$$

$$+ \sum_{n=1}^{N} \sum_{m=1}^{M} \gamma_{nm} (\ln(p_{kn}/C_k))(\ln u_{km}) \leqq 0, k = 1, \cdots, K,$$

(ii) $\sum_{m=1}^{M} \beta_m = 1,$

$$\sum_{m'=1}^{M} \beta_{mm'} = \sum_{m=1}^{M} \gamma_{nm} = 0, m = 1, \cdots, M, n = 1, \cdots, N,$$

(iii) $\alpha_{nn'} = \alpha_{n'n}, n = 1, \cdots, N, n' = 1, \cdots, N,$

$$\beta_{mm'} = \beta_{m'm}, m = 1, \cdots, M, m' = 1, \cdots, M.$$

The maximization over the parameters (α, β, γ), yields the translog cost indirect output distance function frontier for the data $((p/C)^k, u^k), k = 1, \cdots, K$.

The objective constructs the industry or aggregate $ID_o(p/C, u)$ by maximizing the sum of the absolute deviations between $\ln ID_o(p/C, u)$ and $\ln 1$. The first set of constraints (i) ensure that the indirect output distance function is an upper frontier, i.e., $ID_o((p/C)^k, u^k) \leqq 1$ ($\ln ID_o \leqq \ln 1$). The set of constraints (ii) and (iii) impose homogeneity of degree +1 in outputs and symmetry, respectively.

The revenue deflated output shadow prices may now be computed as the derivatives $\frac{\partial ID_o((p/C)^k, u^k)}{\partial u_{km}}, m = 1, \cdots, M, k = 1, \cdots, K$, and using one of the two assumptions (5.2.7) or (5.2.9), we can derive absolute shadow prices. We note that the translog form does not admit zeros in the data.

Before turning to the shadow pricing of inputs, a few observations may prove helpful. In this section we have exploited the relationships among various direct and indirect distance and value functions to derive shadow prices for outputs from the derivative properties of those functions, much like Shephard's lemma. Note that our first example, which used $ID_o(p/C, u)$ to shadow price outputs, we differentiate with respect to the homogeneous argument of $ID_o(p/C, u)$. This yields particularly simple formulations for shadow prices. For the other two examples in which we calculate derivatives of $D_i(u, x)$ and $C(u, p)$, we differentiate with respect to output again, but in these cases, the functions are not homogeneous in output. As a consequence, the shadow price derivations include adjustments for homogeneity through scale elasticity terms. Finally, we note that one may also derive output shadow prices along the general lines developed here based on other functions, for example, the output distance function. One may also derive (shadow) output quantities using similar techniques.

Having completed our discussion of shadow pricing of outputs, we now repeat the exercise for the input side. In order to derive shadow prices for inputs, we recall the duality between the indirect cost function and the indirect input distance function, i.e.,

$$(5.2.19) \quad IC(r/R, p) = \min_x \{px : ID_i(r/R, x) \geqq 1\}$$
$$ID_i(r/R, x) = \min_p \{px : IC(r/R, p) \geqq 1\}.$$

Assume that the revenue indirect distance function is differentiable, then consider the Lagrangian problem associated with the first part of (5.2.19) equals

$$(5.2.20) \quad \wedge = px + \lambda(1 - ID_i(r/R, x)).$$

The first order conditions are

$$(5.2.21) \quad p = \lambda \nabla_x ID_i(r/R, x)$$
$$ID_i(r/R, x) = 1.$$

In Section 4.2 it was proved that at the optimum, the Lagrangian $\lambda = IC(r/R, p)$, thus the first part of (5.2.21) takes the form

$$(5.2.22) \quad p = IC(r/R, p)\nabla_x ID_i(r/R, x).$$

It was shown, see (4.2.18), that the gradient $\nabla_x D_i(r/R, x)$ equals $p^*(r/R, x)$, the cost deflated shadow price vector supporting x. Substituting this observation into (5.2.22) yields

$$(5.2.23) \quad p = IC(r/R, p)p^*(r/R, x).$$

The indirect cost function is homogeneous of degree $+1$ in the input prices we seek, therefore (5.2.23) can only model relative or indirect cost deflated input prices unless we can obtain independent information on $IC(r/R, p)$. To obtain undeflated shadow prices further assumptions are needed. One such assumption is

$$(5.2.24) \quad \textbf{Assumption:} \text{ The firm is not-for-profit.}$$

We interpret this assumption in the same way as Assumption (5.2.7), namely that $IC = R$, i.e., profit is zero. Whenever (5.2.24) is justified, each absolute input price can be computed according to

$$(5.2.25) \quad p_n = R\, p_n^*(r/R, x), n = 1, \cdots, N.$$

where R has been substituted for $IC(r/R, p)$ in (5.2.23).

If the not-for-profit assumption is inappropriate, an alternatie follows in the spirit of Assumption (5.2.9) and states

(5.2.26)   **Assumption:** For $N > 2$, one input price equals its market (observed) price.

Here we suppose that one input price, say p_n^m, equals its market price. From this assumption we can compute indirect costs by

(5.2.27) $IC^* = p_n^m / p_n^*(r/R, x).$

where we have substituted p_n^m in the left hand side of (5.2.23) and rearranged. IC^* is independent of $n = 1, \cdots, N$, thus for $n' \neq n$ we calculate undeflated virtual input prices as

(5.2.28) $p_{n'} = IC^* p_{n'}^*(r/R, x), n' = 1, \cdots, N, n' \neq n,$

where IC^* is from (5.2.27).

The second method for shadow pricing inputs makes use of the relationship between direct and indirect output distance functions. The Lagrangian associated with cost indirect output distance function $ID_o(p/C, u)$, see (5.1.5), is

(5.2.29) $\Lambda = D_o(x, u) + \lambda \left(1 - \dfrac{px}{C} \right).$

The first order conditions are

$$\nabla_x D_o(x, u) = \lambda p / C$$
$$px = C.$$

From these conditions we derive

$$\sum_{n=1}^{N} \frac{x_n \partial D_o(x, u)}{D_o(x, u) \partial x_n} = \frac{\lambda}{D_o(x, u)},$$

where the left hand side is an expression of scale elasticity (see 5.3.3), i.e.,

$\varepsilon_{D_o} = \sum_{n=1}^{N} \frac{\partial D_o(x,u)}{\partial x_n} \frac{x_n}{D_o(x,u)}$, thus

(5.2.30) $\lambda = \varepsilon_{D_o} D_o(x, u),$

and we have the following expression for input prices

(5.2.31) $p_n = \dfrac{C}{\varepsilon_{D_o} D_o(x, u)} \dfrac{\partial D_o(x, u)}{\partial x_n}, n = 1, \cdots, N.$

Whenever information on input and output quantities (x, u) is available, then $D_o(x, u)$ can be calculated, and thus one can obtain $(p_n/C), n = 1, \cdots, N$; to see this, rearrange (5.2.31). In addition if Assumption (5.2.25) or (5.2.26) applies, or if C or some proxy is known, then the absolute input prices $p_n, n = 1, \cdots, N$ can be derived using (5.2.31).

The relationship between the direct and indirect revenue functions is used in our third approach to shadow pricing inputs. The indirect revenue function $IR(p/C, r)$, see (5.1.12), can be written in terms of a Lagrangian expression as

$$(5.2.32) \quad \wedge = R(x, r) + \lambda \left(1 - \frac{px}{C}\right).$$

The corresponding first order conditions are

$$\nabla_x R(x, r) = \lambda p / C$$

$$px = C.$$

From these conditions we may derive

$$\sum_{n=1}^{N} \frac{x_n \partial R(x, r)}{R(x, r) \partial x_n} = \frac{\lambda}{R(x, r)}.$$

The revenue based expression for scale elasticity equals $\varepsilon_R = \sum_{n=1}^{N} \frac{x_n \partial R(x, r)}{R(x, u) \partial x_n}$ (see 5.3.9), thus

$$(5.2.33) \quad \lambda = \varepsilon_R R(x, r),$$

and we obtain the following expression for shadow input prices,

$$(5.2.34) \quad p_n = \frac{C}{\varepsilon_R R(x, r)} \frac{\partial R(x, r)}{\partial x_n}, n = 1, \cdots, N.$$

If information on R, x, and r is available, one may estimate the revenue function $R(x, r)$ and, using (5.2.34), we can calculate $(p_n/C), n = 1, \cdots, N$. If in addition Assumption (5.2.24) or (5.2.24) applies or if minimum cost C or a proxy is known, absolute shadow input prices are obtainable using expression (5.2.34).

Next we give a specific example of how one can compute shadow input prices using the first approach discussed above, based on the revenue indirect input distance function. We assume that $ID_i(r/R, u)$ can be approximated with a translog function, i.e., there are no zeros in the data. Moreover, we assume that there are $k = 1, \cdots, K$ observations on positive revenue deflated output prices $(r/R)^k = (r_{k1}/R_k, \cdots, r_{kM}/R_k)$ and (positive) input quantities $x^k = (x_{k1}, \cdots, x_{kN})$. The

101

parameters of the translog function may be computed as the solution to the following Aigner-Chu type linear programming problem

$$
\min_{(\alpha,\beta,\gamma)} \quad \sum_{k=1}^{K} \left[\alpha_0 + \sum_{m=1}^{M} \alpha_m \ln(r_{km}/R_k) + \sum_{n=1}^{N} \beta_n \ln x_{kn} \right.
$$

$$
+ \frac{1}{2} \sum_{m=1}^{M} \sum_{m'=1}^{M} \alpha_{mm'}(\ln(r_{km}/R_k))(\ln(r_{kn'}/R_k))
$$

$$
+ \frac{1}{2} \sum_{n=1}^{N} \sum_{n'=1}^{N} \beta_{nn'}(\ln x_{kn})(\ln x_{kn'})
$$

$$
\left. + \sum_{m=1}^{M} \sum_{n=1}^{N} \gamma_{mn}(\ln(r_{km}/R_k))(\ln x_{kn}) \right]
$$

subject to (i) $\alpha_0 + \sum_{m=1}^{M} \alpha_m \ln(r_{km}/R_k) + \sum_{n=1}^{N} \beta_n \ln x_{kn}$

$$
+ \frac{1}{2} \sum_{m=1}^{M} \sum_{m'=1}^{M} \alpha_{mm'}(\ln(r_{km}/R_k))(\ln(r_{km'}/R_k))
$$

$$
+ \frac{1}{2} \sum_{n=1}^{N} \sum_{n'=1}^{N} \beta_{nn'}(\ln x_{kn})(\ln x_{kn'})
$$

$$
+ \sum_{m=1}^{M} \sum_{n=1}^{N} \gamma_{mn}(\ln(r_{km}/R_k))(\ln x_{kn}) \geqq 0, k = 1, \cdots, K,
$$

(ii) $\sum_{n=1}^{N} \beta_n = 1, \sum_{n'=1}^{N} \beta_{nn'} = \sum_{n=1}^{N} \gamma_{mn} = 0, m = 1, \cdots, M,$

$n = 1, \cdots, N,$

(iii) $\alpha_{mm'} = \alpha_{m'm}, m = 1, \cdots, M, m' = 1, \cdots, M,$

$\beta_{nn'} = \beta_{n'n}, n = 1, \cdots, N, n' = 1, \cdots, N.$

The minimum over the parameters (α, β, γ) yields the translog frontier of $ID_i(r/R, x)$ for the observed data, the relative shadow prices may be computed as derivatives, and using one of the Assumptions (5.2.25) or (5.2.26), absolute shadow prices may be calculated.

The general remarks made at the end of the discussion of shadow pricing outputs apply here as well. We have not provided an exhaustive menu of how shadow prices of inputs may be derived; for example, one may also derive input shadow prices from the derivatives of the input distance function. Note that shadow pricing formulas are simpler in the case when the associated function is homogeneous in input quantities. One may also, of course, use similar techniques to derive shadow input quantites – Shephard's lemma is the most well-known example.

Finally, in our examples of how to specify some of these functions in order to derive shadow prices, we have used Aigner-Chu type industry frontiers based on linear programming. Since these models are parametric, derivatives are straightforward. They also readily admit inequality constraints, which one may exploit by adding constraints on the signs of the derivatives if, for example, one wishes to shadow price goods and bads in the same model. One may also use stochastic models for shadow pricing as well as the piecewise linear programming models used for efficiency measurement. In the latter case the dual prices from the constraints may be used as a basis for shadow pricing.

5.3 Scale Elasticity and Returns to Scale

Different definitions of scale elasticity have already been introduced in the text. The purpose of this section is to give a complete characterization of the various definitions, and show that they are all equivalent. Moreover, we introduce some global definitions of returns to scale; scale elasticity is only a local measure of returns to scale.

In the case when the technology can be characterized with a scalar-valued production function

(5.3.1) $\quad u = F(x),$

where $F(x) = \max_u \{u : u \in P(x)\}$, scale elasticity is defined as

(5.3.2) $\quad \varepsilon_F = \sum_{n=1}^{N} \frac{\partial F(x) x_n}{\partial x_n F(x)}.$

To illustrate, consider the following simple example. Suppose $F(x) = \sqrt{x}$, then clearly $\varepsilon_F = \frac{1}{2}$.

In order to define scale elasticity for a multioutput technology, let us first consider (5.3.2) in terms of an output distance function. In the scalar output case, $D_o(x, u) = u/F(x)$. Thus (5.3.2) becomes

$$
\begin{aligned}
\sum_{n=1}^{N} \frac{\partial F(x) x_u}{\partial x_n F(x)} &= -\sum_{n=1}^{N} \frac{\partial D_o(x, u)}{\partial x_n} \frac{u x_n}{F(x)(D_o(x, u))^2} \\
&= -\sum_{n=1}^{N} \frac{\partial D_o(x, u)}{\partial x_n} \frac{x_n}{D_o(x, u)}.
\end{aligned}
$$

Thus, in general we define the *Output Distance Function Measure of Scale Elasticity* as

(5.3.3) $\quad \varepsilon_{D_o} = -\sum_{n=1}^{N} \frac{\partial D_o(x, u) x_n}{\partial x_n D_o(x, u)}.$

103

In our simple example, $D_o(x,u) = u/\sqrt{x}$, thus scale elasticity according to (5.3.3) becomes $-(\frac{1}{2}\frac{u}{x^{\frac{3}{2}}}\frac{x\sqrt{x}}{u}) = \frac{1}{2}$. To continue, total differentiation of the output distance function yields

$$\sum_{m=1}^{M}\frac{\partial D_o(x,u)}{\partial u_m}du_m + \sum_{n=1}^{N}\frac{\partial D_o(x,u)}{\partial x_n}dx_n = 0.$$

Define $(prop\ x) = \frac{dx_n}{x_n}$, for $n = 1,\cdots,N$, and $(prop\ u) = \frac{du_m}{u_m}$, $m = 1,\cdots,M$, and introduce these notions into the above expression, then rearranging yields

$$(5.3.4) \qquad \frac{(prop\ u)}{(prop\ x)} = -\frac{\sum_{n=1}^{N}\frac{\partial D_o(x,u)}{\partial x_n}x_n}{\sum_{m=1}^{M}\frac{\partial D_o(x,u)}{\partial u_m}u_m}$$

$$= -\sum_{m=1}^{M}\frac{\partial D_o(x,u)}{\partial x_n}\frac{x_n}{D_o(x,u)},$$

where the last equality holds since the output distance function is homogeneous of degree $+1$ in outputs. Thus we have shown that $\varepsilon_{D_o} = (prop\ u)/(prop\ x)$. Applying the same idea to the input distance function yields

$$(5.3.5) \qquad (prop\ u)/(prop\ x) = -\left(\sum_{m=1}^{M}\frac{\partial D_i(u,x)}{\partial u_m}\frac{u_m}{D_i(u,x)}\right)^{-1} = \varepsilon_{D_i}$$

which we take as the definition of the *Input Distance Function Measure of Scale Elasticity*. This measure then equals the output distance function based measure of scale elasticity.

To define the cost based measure of scale elasticity, suppose that (x,p) are conjugate dual, i.e.,

$$(5.3.6) \qquad px = C(u,p)D_i(u,x).$$

Total cost is assumed to be fixed, i.e., $px = C$. From this expression we obtain $(prop\ x) = -(prop\ p)$, where $(prop\ p) = \frac{dp_n}{p_n}$, $n = 1,\cdots,N$. Now totally differentiate (5.3.6), which yields

$$\sum_{m=1}^{M}C(u,p)\frac{\partial D_i(u,x)}{\partial u_m}du_m + \sum_{n=1}^{N}C(u,p)\frac{\partial D_i(u,x)}{\partial x_n}dx_n +$$

$$\sum_{m=1}^{M}D_i(u,x)\frac{\partial C(u,p)}{\partial u_m}du_m + \sum_{n=1}^{N}D_i(u,x)\frac{\partial C(u,p)}{\partial p_n}dp_n = 0.$$

From this expression we obtain

$$(prop\ u)C(u,p)\sum_{m=1}^{M}\frac{\partial D_i(u,x)}{\partial u_m}\frac{u_m}{} + (prop\ x)C(u,p)D_i(u,x) +$$

$$(prop\ u)D_i(u,x)\sum_{m=1}^{M}\frac{\partial C(u,p)}{\partial u_m}u_m - (prop\ x)C(u,p)D_i(u,x) = 0,$$

where we have exploited the homogeneity of the input distance function in inputs, the homogeneity of the cost function in input prices and the fact that $(prop\ x) = -(prop\ p)$. Dividing through by $(prop\ x)C(u,p)D_i(u,x)$, rearranging and simplifying yields

$$(5.3.7) \qquad \varepsilon_{D_i} = \varepsilon_C = \left(\sum_{m=1}^{M} \frac{\partial C(u,p)u_m}{\partial u_m C(u,p)} \right)^{-1}.$$

The right hand side is taken as the *Cost Based Measure of Scale Elasticity*, ε_C, and hence we have proved that it equals the input distance function based measure of scale elasticity.

Given that (u,r) are conjugate dual, i.e., $ru = D_o(x,u)R(x,r)$, then one can prove that

$$(5.3.8) \qquad \varepsilon_{D_o} = \varepsilon_R,$$

where the *Revenue Based Measure of Scale Elasticity*, ε_R is defined as

$$(5.3.9) \qquad \varepsilon_R = \sum_{n=1}^{N} \frac{\partial R(x,r)}{\partial x_n} \frac{x_n}{R(x,r)}.$$

Thus far we have defined four scale elasticity formulas, all based on the direct technology, and shown that they are equivalent. Next we turn to the indirect technologies and introduce four additional scale elasticity notions. First consider (5.1.7), with $ID_o(p/C,u) = D_o(x,u)$, for $\frac{px}{C} = 1$. Differentiation of this expression gives

$$\sum_{n=1}^{N} \frac{\partial ID_o(p/C,u)}{\partial (p_n/C)} d(p_n/C) + \sum_{m=1}^{M} \frac{\partial ID_o(p/C,u)}{\partial u_m} du_m =$$
$$\sum_{n=1}^{N} \frac{\partial D_o(x,u)}{\partial x_n} dx_n + \sum_{m=1}^{M} \frac{\partial D_o(x,u)}{\partial u_m} du_m$$

and

$$\sum_{n=1}^{N} x_n d\left(\frac{p_n}{C} \right) + \sum_{n=1}^{N} (p_n/C) dx_n = 0.$$

Define $(prop\ p/C) = \frac{d(p_n/C)}{(p_n/C)}$ for all $n = 1, \cdots, N$, then

$$(prop\ p/C) \sum_{n=1}^{N} \frac{\partial ID_o(p/C,u)}{\partial (p_n/C} (p_n/C) + (prop\ u) \sum_{m=1}^{M} \frac{\partial ID_o(p/C,u)}{\partial u_m} u_m =$$
$$(prop\ x) \sum_{n=1}^{N} \frac{\partial D_o(x,u)}{\partial x_n} x_n + (prop\ u) \sum_{m=1}^{M} \frac{\partial D_o(x,u)}{\partial u_m} u_m$$

and

$$(prop\ p/C) = -(prop\ x).$$

105

The last two expressions together with $ID_o(p/C, u) = D_o(x, u)$ yield

$$\sum_{n=1}^{N} \frac{\partial ID_o(p/C, u)(p_n/C)}{\partial(p_n/C)ID_o(p/C, u)} = -\sum_{n=1}^{N} \frac{\partial D_o(x, u) \, x_n}{\partial x_n D_o(x, u)}.$$

Defining the *Cost Indirect Output Based Measure of Scale Elasticity* as

$$(5.3.10) \qquad \varepsilon_{ID_o} = \sum_{n=1}^{N} \frac{\partial ID_o(p/C, u)(p_n/C)}{\partial(p_n/C)ID_o(p/C, u)},$$

we have shown that the direct and indirect output distance function measures of scale elasticity coincide, i.e.,

$$(5.3.11) \qquad \varepsilon_{D_o} = \varepsilon_{ID_o}.$$

The last three measures of scale elasticity are

$$(5.3.12) \qquad \varepsilon_{IC} = -\left(\sum_{m=1}^{M} \frac{\partial IC(r/R, p)(r_m/R)}{\partial(r_m/R)IC(r/R, p)} \right)^{-1},$$

$$(5.3.13) \qquad \varepsilon_{ID_i} = \left(\sum_{m=1}^{M} \frac{\partial ID_i(r/R, x)(r_m/R)}{\partial(r_m/R)ID_i(r/R, x)} \right)^{-1},$$

$$(5.3.14) \qquad \varepsilon_{IR} = -\sum_{n=1}^{N} \frac{\partial IR(p/C, r)(p_n/C)}{\partial(p_n/C)IR(p/C, r)},$$

and they are called the *Cost Indirect*, the *Revenue Indirect Input Based*, and *Revenue Indirect Measures of Scale Elasticity*, respectively. The proofs that the last three formulas are equivalent to their direct counterparts and therefore all of our elasticities are equivalent, are left to the reader. We proceed to global notions of returns to scale.

(5.3.15) **Definition:** The production technology exhibits *Non Increasing Returns to Scale (NIRS)* if $\forall x \in \Re_+^N$ and $\lambda \geq 1, P(\lambda x) \subseteq \lambda P(x)$, it exhibits *Constant Returns to Scale (CRS)* if $\forall x \in \Re_+^N$, and $\lambda > 0, P(\lambda x) = \lambda P(x)$, and it exhibits *Non Decreasing Returns to Scale (NDRS)* if $\forall x \in \Re_+^N$, and $0 < \mu \leq 1, P(\mu x) \subseteq \mu P(x)$.

The definition says that if inputs are proportionally changed, then the technology exhibits *CRS* when outputs are changed in the same proportion. In terms of the output distance function one can prove

(5.3.16) **Proposition:** $P(\lambda x) = \lambda P(x), \lambda > 0$, if and only if $D_o(\lambda x, u) = \lambda^{-1} D_o(x, u)$.

106

Proof: Suppose that $P(\lambda x) = \lambda P(x), \lambda > 0$, then

$$
\begin{aligned}
D_o(\lambda x, u) &= \inf\{\theta : (u/\theta) \in P(\lambda x)\} = \inf\{\theta : (u/\lambda\theta) \in P(x)\} \\
&= \lambda^{-1}\inf\{\delta : (u/\delta) \in P(x)\} = \lambda^{-1}D_o(x, u).
\end{aligned}
$$

Conversely, assume that $D_o(\lambda x, u) = \lambda^{-1}D_o(x, u)$, then

$$
\begin{aligned}
P(\lambda x) &= \{u : D_o(\lambda x, u) \leqq 1\} = \{u : \lambda^{-1}D_o(x, u) \leqq 1\} \\
&= \lambda\{(u/\lambda) : D_o(x, u/\lambda) \leqq 1\} \\
&= \lambda P(x).
\end{aligned}
$$

$$Q.E.D.$$

Now we can associate scale elasticity with *CRS*. Since *CRS* is equivalent to the case in which $D_o(\lambda x, u) = \lambda^{-1}D_o(x, u), \varepsilon_{D_o} = 1$, we say that the technology exhibits constant returns to scale if scale elasticity equals one everywhere. We say that the technology exhibits *Increasing Returns to Scale (IRS)* if it exhibits *NDRS* and not *CRS*. Moreover it exhibits *Decreasing Returns to Scale (DRS)* if it exhibits *NIRS* and not *CRS*. In terms of scale elasticities *IRS* prevails if scale elasticity is everywhere greater than one and *DRS* prevails if scale elasticity is everywhere less than one.

In this section we have showed equivalent ways to describe the scale properties of technology. Similarly, one may derive alternative descriptions of the curvature of technology. We leave these substitution elasticities to the reader, but note in passing that "Morishima" elasticities of substitution may be defined in terms of distance functions.

5.P Problems

(5.P.1)   Under constant returns to scale, show that $\min_{r/R}\{ID_i(r/R, x) : ru \geqq R\} = D_i(u, x)$.

(5.P.2)   Under constant returns to scale, show how the indirect cost function may be derived from the direct revenue function.

(5.P.3)   Sketch the duality theorem between the direct and indirect revenue functions.

(5.P.4)   Sketch the duality theorem between the direct and indirect cost functions.

(5.P.5)   Prove property (5.1.7).

(5.P.6)   Suppose that a single output is produced. Prove Proposition (5.1.6) under these conditions.

(5.P.7)   Use the duality between the revenue function and the output distance function to shadow price outputs.

(5.P.8)   Prove condition (5.3.8).

(5.P.9)   Prove the equivalence between (5.3.12), (5.3.13) and (5.3.14).

(5.P.10) Prove that if the output distance function is homogeneous of degree -1 in inputs, then $\varepsilon_{D_o} = 1$.

Notes

Färe and Primont (1990) have proved the duality between direct and indirect output distance functions under more general conditions than here. The section on shadow pricing is an extension of the work by Färe and Zieschang (1990), which extended work by Färe and Grosskopf (1990). The results on scale elasticity equivalence among the direct distance, cost and revenue functions is due to Färe, Grosskopf and Lovell (1988) and was extended to the indirect case by Fukuyama (1987).

Chapter 6

Efficiency Gauging

6.0 Introduction

In an influential paper, Farrell (1957) derived a decomposition of individual firm performance relative to the goal of cost minimization vis à vis industry best practice. His decomposition partitioned deviation from cost minimization into a price-related component and a technical component, the latter capturing (proportional) deviations from the boundary or isoquant of the best practice technology and the former isolating (as a residual) deviation from minimal cost from choosing the 'wrong' input mix given relative input prices.

This decomposition is illustrated in Figure 6.1. For observation a, producing output level u with inputs x_1^a and x_2^a, total deviation from cost minimization given relative input prices p_1/p_2 can be captured as $0a''/0a$, the ratio of minimal to observed cost (note that minimal cost at a^* equals cost at a'').

Technical efficiency is captured as $0a'/0a$ and price-related or allocative efficiency is measured as the residual $0a''/0a'$, yielding the identity $0a''/0a = (0a'/0a)(0a''/0a')$. As is obvious from the diagram, these Farrell efficiency measures are closely related to the representations of technology introduced in earlier chapters. In particular, the Farrell measure of technical efficiency is the reciprocal of the input distance function. We will exploit this relationship in order to provide alternative means of calculating these efficiency measures.

In this chapter we build on Farrell's work to derive measures of efficiency in an indirect framework, i.e., we judge performance relative to cost or revenue constrained technology instead of relative to the direct technology. This extension should prove useful in judging performance of decisionmaking units such as local governments which operate in a budget restricted environment.

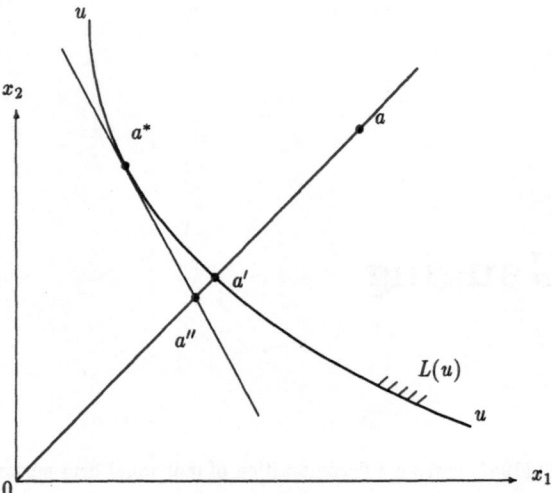

Figure 6.1: Farrell Decomposition

The outline of this chapter is as follows. We begin with the basic Farrell type technical efficiency measures in an indirect framework, i.e., we introduce a revenue indirect input measure of technical efficiency and the cost indirect output measure of technical efficiency. We then derive Farrell type decompositions which include these two technical efficiency measures: (1) we decompose deviations from revenue restricted cost minimization into technical and allocative components and (2) we decompose deviations from cost restricted revenue maximization into technical and allocative components. In the final two sections we introduce a series of linear programming problems which can be used to calculate the efficiency measures introduced in this chapter. These employ the piecewise linear representations of the indirect technology introduced in Chapter 3, as well as a series of parametric linear programming models. A set of problems and notes concludes.

6.1 Indirect Input and Indirect Output Technical Efficiency Measures

In this section we introduce the basic indirect input and indirect output measures of technical efficiency. These illustrate the role of the indirect reference technology and the method of gauging

110

performance relative to the frontier of that technology.

We begin with indirect input efficiency. Suppose that we have information on input quantity vectors $x \in \Re_+^N$ and revenue deflated output prices $(r/R) \in \Re_+^M$, where R represents target revenue. The reference technology relative to which indirect input efficiency is to be judged is the revenue indirect input set

$$(6.1.1) \qquad IL(r/R) = \{x : x \in L(u), \sum_{m=1}^{M} (r_m/R)u_m \geqq 1\},$$

which is the set of all input vectors which can produce target revenue R given output prices r, see (2.4.2).

Now let $(x^k, (r/R)^k)$ be an observed input and revenue deflated output price vector (where $k = 1, \cdots, K$). The technical efficiency of a feasible x^k is measured by

$$(6.1.2) \qquad IF_i((r/R)^k, x^k) = \inf\{\lambda : \lambda x^k \in IL((r/R)^k)\}, k = 1, \cdots, K,$$

and is called the *Revenue Indirect Input Measure of Technical Efficiency*. $IF_i((r/R)^k, x^k)$ gives a measure of the efficiency of input quantity vector x^k in generating revenue R^k at output prices r^k. It computes the ratio of the smallest radial contraction of x^k in $IL((r/R)^k)$ to itself, i.e.,

$$IF_i((r/R)^k, x^k) = \parallel IF_i((r/R)^k, x^k) \cdot x^k \parallel / \parallel x^k \parallel .$$

From the definition of $IF_i((r/R)^k, x^k)$ above it is clear that $IF_i((r/R)^k, x^k) = [ID_i((r/R)^k, x^k)]^{-1}$ since $ID_i((r/R)^k, x^k) = \sup\{\lambda > 0 : (x^k/\lambda) \in IL((r/R)^k)\}$, see (3.2.8). Given that $IL(r/R)$ satisfies IL.1–IL.6, it can be shown that $IF_i((r/R)^k, x^k)$ satisfies properties similar to $(ID_i.1 - ID_i.6)$, accounting for the reciprocal relationship between $IF_i((r/R)^k, x^k)$ and $ID_i((r/R)^k, x^k)$.

Since $IF_i((r/R)^k, x^k)$ is the reciprocal of $ID_i((r/R)^k, x^k)$, it follows that $0 < IF_i((r/R)^k, x^k) \leqq 1$. This is illustrated in Figure 6.2. Consider observed input bundle at a which produces target revenue R at output prices r, i.e., this bundle is an element of $IL(r/R)$. For this input combination $IF_i((r/R)^k, x^k) = 0a'/0a < 1$.

This figure suggests that the value of $IF_i((r/R)^k, x^k)$ attains unity only when the observation under evaluation is an element of the boundary of $IL(r/R)$. More precisely, we can state the following

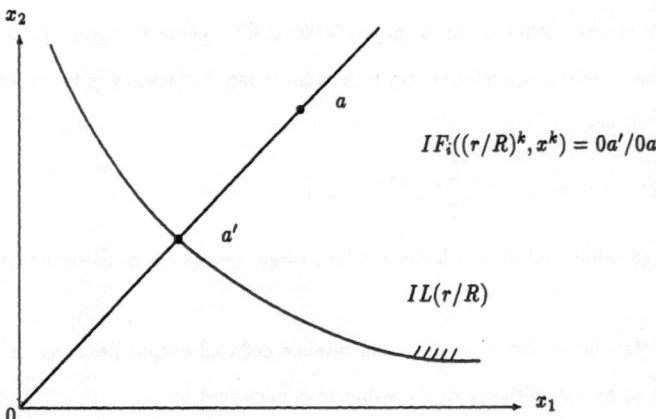

$$IF_i((r/R)^k, x^k) = 0a'/0a$$

$$IL(r/R)$$

Figure 6.2: Revenue Indirect Input Measure of Technical Efficiency

(6.1.3)   **Proposition:** For $x^k \in IL((r/R)^k)$, $IF_i((r/R)^k, x^k) = 1$ if and only if
$x^k \in IsoqIL((r/R)^k)$, where
$$IsoqIL(r/R) = \{x \in IL(r/R) : \lambda x \notin IL(r/R), 0 < \lambda < 1\}.$$

Proof: Let $IF_i((r/R)^k, x^k) = 1$, and suppose that $x^k \notin IsoqIL((r/R)^k)$. Since $x^k \in IL((r/R)^k)$,
then there exists a $\lambda \in (0,1)$ such that $\lambda x^k \in IL((r/R)^k)$, contradicting $IF_i((r/R)^k, x^k) = 1$. The
converse may also be proved by contradiction and is left to the reader.

Q.E.D.

If more specific assumptions are made as to the scale and disposability properties of the
reference technology $IL(r/R)$, a decomposition of $IF_i((r/R)^k, x^k)$ into scale, congestion and
purely technical efficiency components may be undertaken. In Section 6.3 we provide an example
of such a decomposition when efficiency is measured using linear programming techniques based
on specifications of the reference technology as piecewise linear. Such technologies were
introduced in Chapter 3.

As mentioned above, $IF_i((r/R)^k, x^k)$ is the reciprocal of the revenue indirect input distance
function. Similarly, we can define an indirect measure of technical efficiency which is reciprocal to
the cost indirect output distance function. In this case the reference technology is the cost

112

indirect output set

(6.1.4) $IP(p/C) = \{u : u \in P(x), \sum_{n=1}^{N} p_n x_n \leqq C\}$

where C is a prespecified positive budget or cost and $p \in \Re_+^N$ is a vector of input prices, assumed to be given.

Now let $(u^k, (p/C)^k)$ be an observed output quantity and cost deflated input price vector for firm k. The efficiency of feasible u^k is measured by

(6.1.5) $IF_o((p/C)^k, u^k) = \sup\{\theta : \theta u^k \in IP((p/C)^k)\}, k = 1, \cdots, K,$

and is called the *Cost Indirect Output Measure of Technical Efficiency*. $IF_o((p/C)^k, u^k)$ gives a measure of the efficiency of firm k in transforming its budget C^k at prices p^k into outputs. It computes the ratio of the largest radial expansion of u^k in $IP((p/C)^k)$ to itself, i.e.,

$IF_o((p/C)^k, u^k) = \| IF_o((p/C)^k, u^k)u^k \| / \| u^k \|.$

The cost indirect output measure of technical efficiency is illustrated in Figure 6.3. Consider observed output bundle at a which is generated from budget C at input prices p, i.e., a is an element of $IP(p/C)$. For this observation, $IF_o((p/C)^a, u^a) = 0a'/0a > 1$. That is, firm a could have increased its relative outputs by $0a'/0a - 1$ percent if it had been operating on the frontier of $IP(p/C)$, given its output mix.

From this diagram, it is clear that $IF_o((p/C)^k, u^k)$ is closely related to the cost indirect output distance function. Since $ID_o((p/C)^k, u^k) = \inf\{\theta > 0 : (u^k/\theta) \in IP((p/C)^k)\}$, it is clear that $ID_o((p/C)^k, u^k)$ is the reciprocal of $IF_o((r/R)^k, u^k)$.

Given the reciprocal relationship between $ID_o((p/C)^k, u^k)$ and $IF_o((p/C)^k, u^k)$, it follows that $IF_o((p/C)^k, u^k)$ inherits properties from the parent technology analogous to $(ID_o.1 - ID_o.6)$, see Section 3.1. As is clear from Figure 6.3, $IF_o((p/C)^k, u^k)$ achieves its lower bound of unity when u^k is on the boundary of $IP((p/C)^k)$. More precisely

(6.1.6)   **Proposition:** For $u^k \in IP((p/C)^k)$, $IF_o((p/C)^k, u^k) = 1$ if and only if

$u^k \in IsoqIP((p/C)^k)$, where $IsoqIP(p/C) = \{u \in IP(p/C) : \theta u \notin IP(p/C), \theta > 1\}$.

The proof is similar to that for Proposition (6.1.3) and is omitted.

In the discussion above, we have not specified the properties satisfied by $IP(p/C)$ with respect to scale properties or disposability. As it turns out, we can decompose $IF_o((p/C)^k, u^k)$ into

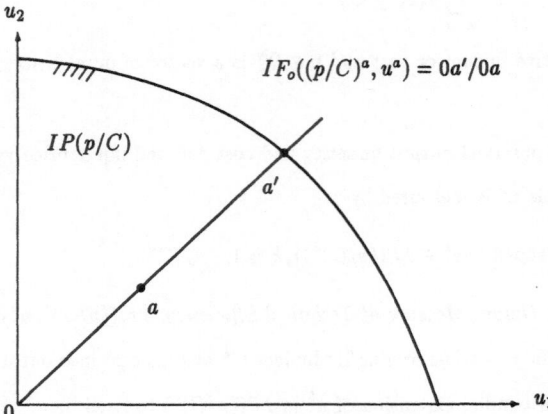

Figure 6.3: Cost Indirect Output Measure of Technical Efficiency

deviations from efficient scale and purely technical efficiency by systematically varying assumptions on the reference technology. This is demonstrated in Section 6.3, where we introduce linear programming methods to calculate the efficiency measures introduced in this chapter. In that section we use the piecewise linear representations of $IP(p/C)$ introduced in Chapter 3.

6.2   Decomposition of Indirect Efficiency

In this section we derive indirect analogs of the Farrell decomposition of efficiency described in the introduction. We begin with a decomposition of deviations from revenue indirect cost minimization and then turn to a decomposition of deviations from cost indirect revenue maximization.

Our first decomposition is based on the ability of a firm to minimize costs when they are revenue restricted. Specifically, the firm's goal is to minimize revenue indirect costs

$$(6.2.1) \qquad IC((r/R)^k, p^k) \;=\; \inf_x \{p^k x : x \in IL((r/R)^k)\}$$
$$= \inf_x \{p^k x : x \in L(u), r^k u \geqq R^k\}.$$

If we assume that p is positive and $IL(r/R)$ nonempty, then infimum in (6.2.1) can be replaced

114

with minimum (see Section 4.2). In general we have $p^k x \geqq IC((r/R)^k, p^k)$, which suggests the following measure of overall performance for firms which are revenue restricted.

(6.2.2) $IO_i((r/R)^k, p^k, x^k) = IC((r/R)^k, p^k)/p^k x^k, k = 1, \cdots, K,$

is called the *Revenue Indirect Input Cost Efficiency Measure*. It calculates the ratio of revenue constrained minimal cost to observed cost and takes values between zero and one. It attains its maximum value of unity when the firm minimizes indirect cost, i.e., when the firm attains the minimum in (6.2.1).

(6.2.3)   **Proposition**: With $p^k > 0, IO_i((r/R)^k, p^k, x^k) = 1 \Longleftrightarrow x^k$ solves (6.2.1).

Proof: Suppose $IO_i((r/R)^k, p^k, x^k) = 1$, but x^k does not minimize costs given p^k and $IL((r/R)^k)$. Then by Definition (6.2.2), $IO_i((r/R)^k, p^k, x^k) < 1$, a contradiction. For the converse, let x^k solve (6.2.1) but assume that $IO_i((r/R)^k, p^k, x^k) < 1$. Given our assumption that $p^k > 0$ (and $IL((r/R)^k)$ is closed and nonempty), a minimum is achieved in (6.2.1) and $p^k x^k = IC((r/R)^k, p^k)$, a contradiction.

$$Q.E.D.$$

Next we wish to decompose $IO_i((r/R)^k, p^k, x^k)$ into allocative and technical components. From Section 6.1 we have a measure of revenue indirect technical efficiency, namely $IF_i((r/R)^k, x^k)$, which is independent of input prices. This allows us to define a measure of allocative efficiency as the residual deviation from revenue indirect input cost efficiency.

(6.2.4)   **Definition:** $IA_i((r/R)^k, p^k, x^k) = \frac{IO_i((r/R)^k, p^k, x^k)}{IF_i((r/R)^k, x^k)}, k = 1, \cdots, K$ is called the *Revenue Indirect Allocative Efficiency Measure*.

This measure provides an indication of how the chosen input mix at x^k deviates from the optimal mix, given input prices p^k. We note that $IA_i((r/R)^k, p^k, x^k)$ takes on values between zero and unity. To see that it will be less than or equal to one, rewrite (6.2.4)

(6.2.5) $IA_i((r/R)^k, p^k, x^k) = \frac{IC((r/R)^k, p^k)}{p^k x^k (IF_i((r/R)^k, x^k))}.$

From Proposition (6.1.3) and the definition of $IF_i((r/R)^k, x^k), x^k IF_i((r/R)^k, x^k) \in IsoqIL((r/R)^k)$, thus the denominator in (6.2.5) is greater than or equal to $IC((r/R)^k, p^k)$.

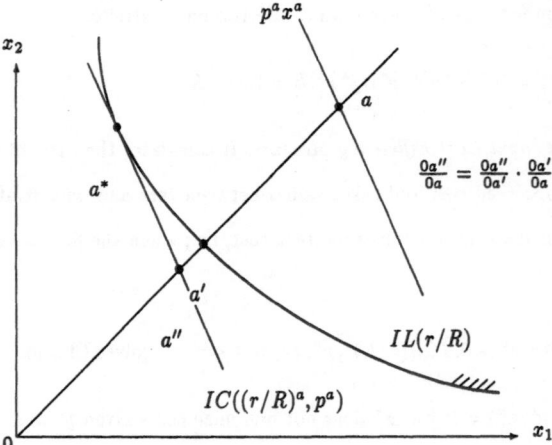

Figure 6.4: Decomposition of Revenue Indirect Input Cost Efficiency

We may now state our decomposition of revenue indirect input cost efficiency for observations $k = 1, \cdots, K$ as

$$(6.2.6) \qquad IO_i((r/R)^k, p^k, x^k) = IA_i((r/R)^k, p^k, x^k) \cdot IF_i((r/R)^k, x^k),$$

i.e., deviations from revenue indirect minimal cost are due to departures from optimal input mix given prices p^k, and deviations from technical efficiency. This is illustrated in Figure 6.4. Given the observed input bundle x^a at a and relative input prices implied by the hyperplane supporting a^*, the ratio of minimal to observed indirect cost is equivalent to $0a''/0a$ (note that cost at a^* equals cost at a''). This can be decomposed into a technical component: $0a'/0a$ and a residual allocative component: $0a''/0a'$.

As discussed in the introduction, these indirect measures of efficiency are appropriate for assessing performance of decisionmaking units facing revenue or budget restrictions. Here the firm faces a revenue requirement and is judged on its ability to meet that revenue target with minimal resource use. In calculating $IO_i((r/R)^k, p^k, x^k)$ we compare minimal costs required to meet the revenue target with observed costs. One may also wish to compare these costs to "benefits", where benefits are measured based on achieved revenues. That is, one could calculate an *Expense Effectiveness Ratio* as $(R^k/IC((r/R)^k, p^k))$ for each observation $k = 1, \cdots, K$. If these individual

116

observations are alternative potential public projects, for example, requiring that projects have an expense effectiveness ratio greater than unity could be a type of project evaluation criterion.

Before turning to the decomposition of cost indirect efficiency measures, we note that, if instead of revenue deflated output prices r/R, only information on output quantities u is available, a series of measures analogous to those developed above (substituting u^k for $(r/R)^k$) can be developed. The resulting measures would be identical to the original Farrell input based measures discussed in the introduction.

In parallel to the discussion above, we can derive a set of efficiency measures and a decomposition of those measures based on the goal of cost indirect revenue maximization. Specifically, recall from (4.1.1) the definition of the cost indirect revenue function

$$
(6.2.7) \qquad
\begin{aligned}
IR(p/C, r) &= \sup_{u}\{ru : u \in IP(p/C)\} \\
&= \sup_{u}\{ru : u \in P(x), px \leqq C\},
\end{aligned}
$$

where C refers to target cost, and $IP(p/C)$ is the cost restricted output set. Here we assume that cost deflated input prices are positive, $IP(p/C)$ is nonempty and output prices nonnegative, thus the supremum is achieved in (6.2.7) and may be replaced with maximum. Clearly $ru \leqq IR(p/C, r)$ which suggests a measure of overall performance when a firm is budget restricted. For each observations $k = 1, \cdots, K$

$$
(6.2.8) \qquad IO_o((p/C)^k, r^k, u^k) = IR((p/C)^k, r^k)/r^k u^k,
$$

is called the *Cost Indirect Output Measure of Revenue Efficiency*. It calculates the ratio of budget constrained maximal revenue to observed revenue for firm k. If observation k meets the budget constraint, then $IO_o((p/C)^k, r^k, u^k) \geqq 1$. This measure achieves a value of unity when the firm to be evaluated maximizes indirect revenue, i.e., when the firm achieves the maximum in (6.2.7).

(6.2.9) **Proposition:** With $r^k \geq 0, IO_o((p/C)^k, r^k, u^k) = 1$ if and only if u^k solves (6.2.7).

The proof parallels that of (6.2.3) and is left to the reader.

Next we decompose $IO_o((p/C)^k, r^k, u^k)$ into mutually exclusive and exhaustive allocative and technical components. From Section (6.1) we have a measure of cost indirect technical efficiency, namely $IF_o((p/C)^k, r^k, u^k)$. We can now define a measure of allocative efficiency as the residual between $IO_o((p/C)^k, r^k, u^k)$ and $IF_o((p/C)^k, u^k)$.

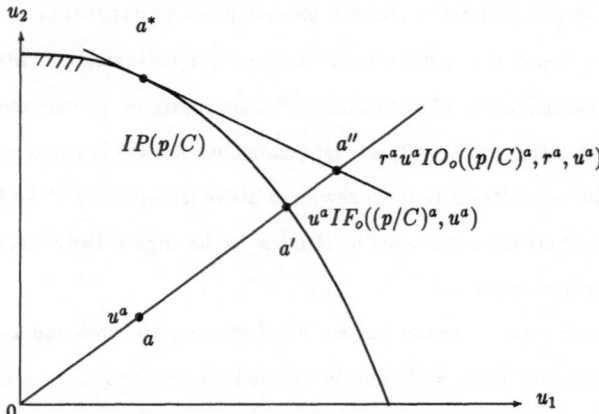

Figure 6.5: Decomposition of Cost Indirect Output Revenue Efficiency

(6.2.10)   **Definition:** $IA_o((p/C)^k, r^k, u^k) = \frac{IO_o((p/C)^k, r^k, u^k)}{IF_o((p/C)^k, u^k)}, k = 1, \cdots, K$ is called the *Cost Indirect Allocative Efficiency Measure.*

This measure provides an indication of how the chosen output mix u^k deviates from the optimal mix, given output prices r^k and technology $IP((p/C)^k)$.

This measure takes on values greater than or equal to unity. This is intuitively clear from the definition of $IO_o((p/C)^k, r^k, u^k)$. Rewrite (6.2.10) as $IA_o((p/C)^k, r^k, u^k) = \frac{IR((p/C)^k, r^k)}{r^k u^k IF_o((p/C)^k, u^k)}$, i.e., allocative efficiency measures the ratio of maximal cost restricted revenue to the revenue of a technically efficient point $r^k u^k IF_o((p/C)^k, u^k)$.

This leads us to a decomposition of cost indirect output revenue efficiency derived by rearranging (6.2.10) for each observation $k = 1, \cdots, K$ as

(6.2.11) $IO_o((p/C)^k, r^k, u^k) = IA_o((p/C)^k, r^k, u^k) \cdot IF_o((p/C)^k, u^k).$

This is illustrated in Figure 6.5.

Given observed output bundle u^a at point a, observed prices r^a and technology $IP(p/C)$, firm a maximizes cost indirect revenues at a^*, with revenues $r^a u^a IO_o((p/C)^a), r^a, u^a)$, where $IO_o((p/C)^a, r^a, u^a) = 0a''/0a$. Our decomposition in (6.2.11) is confirmed for point a since $0a''/0a = (0a''/0a')(0a'/0a)$.

As discussed in the introduction these cost indirect measures of efficiency are appropriate for gauging performance when an observation is budget restricted. Public sector activity, for example, may be an appropriate arena for application of these performance measures. In that context it may also be of interest to augment the cost indirect revenue efficiency measure $IO_o((p/C)^k, r^k, u^k)$ above, which compares maximal budget constrained revenue to observed revenue, with a measure that compares maximal budget constrained revenues with the budget. We could calculate a *Benefit Effectiveness Ratio* as $(IR((p/C)^k, r^k)/C^k)$ for each observation $k = 1, \cdots, K$. If these are alternative projects, they could be ranked by this measure and rejected if $(IR((p/C)^k, r^k)/C^k < 1)$. Note that this measure compares projects with respect to their potential for converting budgets into revenues since technical and allocative inefficiency are eliminated from the numerator.

Our measures of cost indirect revenue efficiency require information on input prices p^k and budgets $C^k, k = 1, \cdots, K$. If such information is not available (or firms are not budget constrained), one may substitute x^k for $(p/C)^k, k = 1, \cdots, K$ and develop a series of budget unrestricted output efficiency measures analogous to the budget restricted measures developed here. These would be equivalent to what is called Farrell output based measure of efficiency.

6.3 Calculating Efficiency: Nonparametric Approach

In this section we use the piecewise linear models of technology developed in Chapter 3 to describe the reference sets in Section 6.2 as linear constraints, and develop linear programming problems to calculate our indirect efficiency measures. This approach has the advantage of minimal data requirements, simplicity and considerable flexibility. No explicit functional form is imposed on the data, rather the data are 'enveloped', using programming methods to form the best practice frontier.

Before introducing the programming problems, recall from Section 3.3 that the piecewise linear reference technologies employ a vector of intensity variables $z = (z_1, \cdots, z_K)$, one intensity variable for each activity or observation $k = 1, \cdots, K$. These serve to construct the linear segments between activities or observations to form the technology and its boundary. In this section we index observations by $k = 1, \cdots, K$ and assume that inputs $x^k \in \Re_+^N$ and outputs $u^k \in \Re_+^M$ satisfy the restrictions specified in (3.3.1) which ensure that feasible solutions to our programming problems exist. We also assume that firms face input prices $p^k \in \Re_+^N$, output prices

119

$r^k \in \Re_+^M$, target costs $C^k > 0$ and revenues $R^k > 0$. We note that if all firms face the same (exogenous) prices, the k superscripts may be dropped from prices.

We begin with the revenue indirect input measure of technical efficiency, defined as $IF_i((r/R)^k, x^k) = \inf\{\lambda : \lambda x^k \in IL((r/R)^k)\}, k = 1, \cdots, K$ (see (6.1.2)). This can be specified as the solution to a linear programming problem by using the piecewise linear formulation of $IL(r/R)$ from (3.3.12). Specifically, for each $k = 1, \cdots, K$

$$(6.3.1) \quad IF_i((r/R)^k, x^k \mid CRS) = \min_{(\lambda, z, u)} \lambda$$

$$s.t. \quad \sum_{k=1}^{K} z_k u_{km} \geqq u_m, m = 1, \cdots, M,$$

$$\sum_{k=1}^{K} z_k x_{kn} \leqq \lambda x_{kn}, n = 1, \cdots, N,$$

$$z_k \geqq 0, k = 1, \cdots, K,$$

$$\sum_{m=1}^{M} r_{km} u_m \geqq R_k.$$

Given our assumptions on data and technology, the infimum in (6.1.2) can be taken as a minimum. The constraints have been rearranged slightly to more closely approximate canonical form. Note, however that u_m and λ are choice variables and strictly speaking would appear on the left hand side in canonical form. Note that we have added a "CRS" term in $IF_i((r/R)^k, x^k \mid CRS)$, which stands for constant returns to scale. As discussed in Section 3.3, the restriction that $z_k \geqq 0, k = 1, \cdots, K$ imposes constant returns to scale on the technology. We will eventually relax that assumption in order to isolate scale inefficiency.

Next we turn to the calculation of revenue indirect input cost efficiency, $IO_i((r/R)^k, p^k, x^k)$ which requires the calculation of revenue indirect minimal cost, i.e., we minimize costs subject to the piecewise linear representation of $IL(r/R)$. Specifically, for $k = 1, \cdots, K$

$$(6.3.2) \quad IC((r/R)^k, p^k) = \min_{(z, u, x)} \sum_{n=1}^{N} p_{kn} x_n$$

$$s.t. \quad \sum_{k=1}^{K} z_k u_{km} \geqq u_m, m = 1, \cdots, M,$$

$$\sum_{k=1}^{K} z_k x_{kn} \leqq x_n, n = 1, \cdots, N,$$

$$z_k \geqq 0, k = 1, \cdots, K,$$

120

$$\sum_{m=1}^{M} r_{km} u_m \geqq R_k.$$

Note that in contrast to (6.3.1), x becomes a choice variable here. The solution to (6.3.2) can then be used to construct $IO_i((r/R)^k, p^k, x^k) = IC((r/R)^k, p^k)/p^k x^k, k = 1, \cdots, K$, where $p^k x^k$ is observed cost.

Allocative efficiency can be calculated using the definition of $IA_i((r/R)^k, p^k, x^k) = IO_i((r/R)^k, p^k, x^k)/IF_i((r/R)^k, x^k \mid CRS), k = 1, \cdots, K$.

As mentioned above, and discussed in detail in Section 3.3, we can vary the scale properties of the technology by varying the restrictions on the z vector of intensity variables. So far we have imposed the restriction $z_k \geqq 0, k = 1, \cdots, K$, which imposes constant returns to scale, i.e., we have defined $IF_i((r/R)^k, x^k \mid CRS), IO_i((r/R)^k, p^k, x^k)$ and $IA_i((r/R)^k, p^k, x^k)$ relative to $IL(r/R)$ satisfying constant returns to scale. In order to provide a decomposition of $IF_i((r/R)^k, x^k \mid CRS)$ into purely technical and scale related components, we vary the restrictions on the intensity variable.

If we restrict the intensity vector to sum to unity, the piecewise linear reference technology is no longer restricted to satisfy constant returns, but rather may exhibit increasing, constant or decreasing returns to scale, which we shall refer to as variable returns to scale, or VRS. Thus we may calculate a VRS *Measure of Cost Indirect Input Technical Efficiency* as

$$(6.3.3) \quad IF_i((r/R)^k, x^k \mid VRS) = \min_{(\lambda, z, u)} \lambda$$

$$s.t. \quad \sum_{k=1}^{K} z_k u_{km} \geqq u_m, m = 1, \cdots, M,$$

$$\sum_{k=1}^{K} z_k x_{kn} \leqq \lambda x_{kn}, n = 1, \cdots, N,$$

$$z_k \geqq 0, k = 1, \cdots, K,$$

$$\sum_{m=1}^{M} r_{km} u_m \geqq R_k$$

$$\sum_{k=1}^{K} z_k = 1.$$

We may now define a measure of *Revenue Indirect Scale Efficiency* as

$$(6.3.4) \quad IS_i((r/R)^k, x^k) = IF_i((r/R)^k, x^k \mid CRS)/IF_i((r/R)^k, x^k \mid VRS).$$

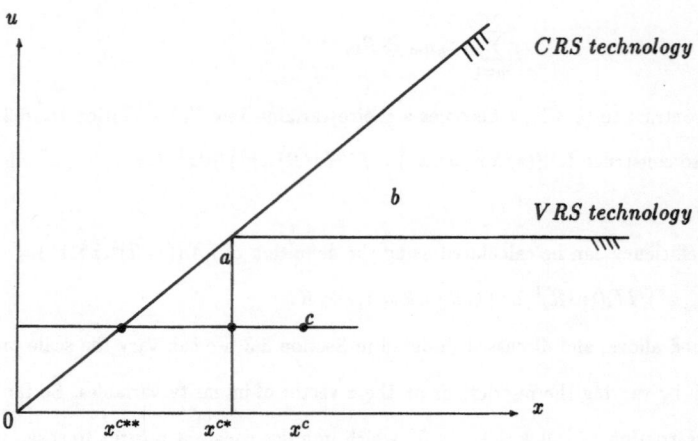

Figure 6.6: Indirect Scale Efficiency

This measure captures the degree to which an observation k deviates from constant returns to scale, where "purely" technical efficiency has been removed. It is illustrated in Figure 6.6 for the case of a single input and output. Suppose there are three observations, a, b and c. The technology satisfying constant returns to scale (CRS) is constructed as the cone through a. The variable returns technology (VRS) is bounded by $x^{c*}ab$ and the horizontal extension from b. Consider c. In this diagram, $x^{c*} = IF_i((r/R)^c, x^c \mid VRS) \cdot x^c$, and $x^{c**} = IF_i((r/R)^c, x^c \mid CRS) \cdot x^c$, thus $IS_i((r/R)^c, x^c) = 0x^{c**}/0x^{c*}$. From this diagram, it is clear that $IS_i((r/R)^k, x^k) \leqq 1$, and that it measures the "x-distance", between the variable and constant returns to scale frontiers.

For observation c it is obvious that $IS_i((r/R)^c, x^c) < 1$ and deviation from constant returns to scale for point c is due to operation in an area of increasing returns (the segment $x^{c*}a$ of the VRS technology). Generally speaking, however, $IS_i((r/R)^k, x^k) < 1$ could also be due to decreasing returns. To determine whether $IS_i((r/R)^k, x^k) < 1$ is due to operation in a range of increasing or decreasing returns, one can calculate

$$(6.3.5) \quad IF_i((r/R)^k, x^k \mid NIRS) = \min_{(\lambda, z, u)} \lambda$$

$$s.t. \quad \sum_{k=1}^{K} z_k u_{km} \geqq u_m, m = 1, \cdots, M,$$

122

$$\sum_{k=1}^{K} z_k x_{kn} \leqq \lambda x_{kn} n = 1, \cdots, N,$$

$$z_k \geqq 0, k = 1, \cdots, K,$$

$$\sum_{m=1}^{M} r_{km} u_m \geqq R_k,$$

$$\sum_{k=1}^{K} z_k \leqq 1.$$

(6.3.5) differs from (6.3.4) only in the restriction on the intensity variables. In (6.3.5) they are restricted to sum to less than or equal to unity, which imposes nonincreasing returns to scale (denoted by $NIRS$ in (6.3.5)). In Figure 6.6, the $NIRS$ technology would be bounded by $0ab$ and the horizontal extension from b.

By comparing $IF_i((r/R)^k, x^k \mid NIRS)$ and $IF_i((r/R)^k, x^k \mid VRS)$, one can determine whether $IS_i((r/R)^k, x^k) < 1$ is due to increasing or decreasing returns. If $IS_i((r/R)^k, x^k) < 1$ and $IF_i((r/R)^k, x^k \mid VRS) = IF_i((r/R)^k, x^k \mid NIRS)$, then scale inefficiency is due to decreasing returns. Otherwise scale inefficiency is due to increasing returns.

We may now use our definition of scale elasticity to derive a composition of $IF_i((r/R)^k, x^k \mid CRS)$, i.e.,

(6.3.6) $\qquad IF_i((r/R)^k, x^k \mid CRS) = IS_i((r/R)^k, x^k) \cdot IF_i((r/R)^k, x^k \mid VRS).$

Furthermore, this allows us to augment our original decomposition of $IO_i((r/R)^k, p^k, x^k)$ to include scale efficiency, i.e.,

(6.3.7) $\qquad IO_i((r/R)^k, p^k, x^k) = IA_i((r/R)^k, p^k, x^k) \cdot IS_i((r/R)^k, x^k) \cdot IF_i((r/R)^k, x^k \mid VRS).$

Next we turn to the calculation of our cost indirect output efficiency measures. We begin with the calculation of the CRS cost indirect output measure of technical efficiency for observations $k = 1, \cdots, K$ which is calculated as the solution to the following linear programming problem

(6.3.8) $\qquad IF_o((p/C)^k, u^k \mid CRS) \quad = \quad \max_{(\theta, z, x)} \theta$

$$s.t. \quad \sum_{k=1}^{K} z_k u_{km} \geqq \theta u_{km}, m = 1, \cdots, M,$$

$$\sum_{k=1}^{K} z_k x_{kn} \leqq x_n, n = 1, \cdots, N,$$

$$z_k \geqq 0, k = 1, \cdots, K,$$

$$\sum_{n=1}^{N} p_{kn}x_n \leqq C_k.$$

Note that our only restriction on the intensity variables is that they be nonnegative, therefore the technology satisfies constant returns to scale.

In order to calculate $IO_o((p/C)^k, r^k, u^k)$, we first need to find maximal cost indirect revenue

$$(6.3.9) \quad IR((p/C)^k, r^k) = \max_{(z,x,u)} \sum_{m=1}^{M} r_{km}u_m$$

$$s.t. \quad \sum_{k=1}^{K} z_k u_{km} \geqq u_m, m = 1, \cdots, M,$$

$$\sum_{k=1}^{K} z_k x_{kn} \leqq x_n, n = 1, \cdots, N,$$

$$z_k \geqq 0, k = 1, \cdots, K,$$

$$\sum_{n=1}^{N} p_{kn}x_n \leqq C_k.$$

The solution to (6.3.9) can be used to construct

$IO_o((p/C)^k, r^k, u^k) = IR((p/C)^k, r^k)/r^k u^k, k = 1, \cdots, K$, where $r^k u^k$ is observed revenue.

Next we can calculate allocative efficiency using the definition

$$(6.3.10) \quad IA_o((p/C)^k, r^k, u^k) = IO_o((p/C)^k, r^k, u^k)/IF_o((p/C)^k, u^k \mid CRS).$$

In parallel to the discussion of the indirect input efficiency measures, we can decompose $IF_o((p/C)^k, u^k \mid CRS)$ into purely technical and scale related components. First calculate purely technical efficiency as a *VRS Measure of Cost Indirect Technical Efficiency* for each observation $k = 1, \cdots, K$ as

$$(6.3.11) \quad IF_o((p/C)^k, u^k \mid VRS) = \max_{(\theta,z,x)} \theta$$

$$s.t. \quad \sum_{k=1}^{K} z_k u_{km} \geqq \theta u_{km}, m = 1, \cdots, M,$$

$$\sum_{k=1}^{K} z_k x_{kn} \leqq x_n, n = 1, \cdots, N,$$

$$z_k \geqq 0, k = 1, \cdots, K,$$

$$\sum_{n=1}^{N} p_{kn}x_n \leqq C_k,$$

$$\sum_{k=1}^{K} z_k = 1.$$

(6.3.11) differs from $IF_o((p/C)^k, u^k \mid CRS)$ only in the additional constraint $\sum_{n=1}^{K} z_k = 1$ which allows the technology to exhibit variable returns to scale (denoted in (6.3.11) as VRS). We can now calculate *Cost Indirect Scale Efficiency* as

$$(6.3.12) \qquad IS_o((p/C)^k, u^k) = IF_o((p/C)^k, u^k \mid CRS)/IF_o((p/C)^k, u^k \mid VRS),$$

for each observation $k = 1, \cdots, K$.

If $IS_o((p/C), u^k) < 1$, observation k is not operating at constant returns to scale (CRS), but rather at increasing returns (IRS) or decreasing returns (DRS). To distinguish between IRS and DRS when a firm is scale inefficient, we need to calculate $IF_o((p/C)^k, u^k \mid NIRS)$, which is calculated according to (6.3.11) except that the intensity variables are restricted to sum to less than or equal to unity $(\sum_{k=1}^{K} z_k \leqq 1)$, which imposes nonincreasing returns to scale $(NIRS)$. If $IS_o((p/C)^k, u^k) < 1$ and $IF_o((p/C)^k, u^k \mid VRS) = IF_o((p/C)^k, u^k \mid NIRS)$ then scale inefficiency is due to DRS. If they are not equal, scale inefficiency is due to firm k's operation at a point of IRS.

Note that our original Farrell type decomposition

$$(6.3.13) \qquad IO_o((p/C)^k, r^k, u^k) = IA_o((p/C)^k, r^k, u^k) \cdot IF_o((p/C)^k, u^k),$$

may now be augmented to include scale efficiency along the lines described in (6.3.6) and (6.3.7).

6.4 Calculating Efficiency: Parametric Approach

In this section we provide examples of a parametric frontier approach to calculating efficiency. Here we focus on Aigner-Chu type models. These have been called deterministic, parametric models, i.e., those based on programming models which have parametric functional forms. One may also specify what have been called parametric stochastic frontier models using composed error terms to capture inefficiency.

In earlier chapters we introduced two parameterizations of indirect distance, cost and revenue functions: a translog and a variation of generalized Leontief. Here we focus on the translog form and leave the generalized Leontief form to the reader. We begin with the decomposition of revenue indirect input efficiency.

We first calculate revenue indirect input technical efficiency. Recall that Farrell type technical efficiency measures are the reciprocals of the corresponding distance functions. In this case

$IF_i((r/R)^k, x^k)$ is the reciprocal of $ID_i((r/R)^k, x^k)$. Accordingly, we can modify the translog

formulation of the revenue indirect input distance function (see (5.2.35)) to the following problem

$$
(6.4.1) \quad \max_{(\alpha,\beta,\gamma)} \sum_{k=1}^{K} \left[\alpha_o + \sum_{m=1}^{M} \alpha_m \ln(r_{km}/R_k) + \sum_{n=1}^{N} \beta_n \ln x_{kn} \right.
$$

$$
+ \frac{1}{2} \sum_{m=1}^{M} \sum_{m'}^{M} \alpha_{mm'}(\ln(r_{km}/R_k))(\ln(r_{km'}/R_k))
$$

$$
\left. + \frac{1}{2} \sum_{n=1}^{N} \sum_{n'}^{N} \beta_{nn'}(\ln x_{kn})(\ln x_{kn'}) + \sum_{m=1}^{M} \sum_{n=1}^{N} \gamma_{mn}(\ln(r_{km}/R_k))(\ln x_{kn}) \right]
$$

$$
\text{s.t.} \quad \text{(i)} \quad \alpha_o + \sum_{m=1}^{M} \alpha_m \ln(r_{km}/R_k) + \sum_{n=1}^{N} \beta_n \ln x_{kn}
$$

$$
+ \frac{1}{2} \sum_{m=1}^{M} \sum_{m'=1}^{M} \alpha_{mm'}(\ln(r_{km}/R_k))(\ln(r_{km'}/R_k))
$$

$$
+ \frac{1}{2} \sum_{n=1}^{N} \sum_{n'=1}^{N} \beta_{nn'}(\ln x_{kn})(\ln x_{kn'})
$$

$$
+ \sum_{m=1}^{M} \sum_{n=1}^{N} \gamma_{mn}(\ln(r_{km}/R_k))(\ln x_{kn}) \leqq 0, k = 1, \cdots, K,
$$

$$
\text{(ii)} \quad \sum_{n=1}^{N} \beta_n = 1,
$$

$$
\sum_{n=1}^{N} \beta_{nn'} = \sum_{m=1}^{M} \gamma_{mn} = 0, m = 1, \cdots, M, n = 1, \cdots, N,
$$

$$
\text{(iii)} \quad \alpha_{mm'} = \alpha_{m'm}, m = 1, \cdots, M, m' = 1, \cdots, M,
$$

$$
\beta_{nn'} = \beta_{n'n}, n = 1, \cdots, N, n' = 1, \cdots, N.
$$

This formulation yields estimates of the industry frontier revenue indirect technology (from the

objective), and allows us to calculate $IF_i((r/R)^k, x^k \mid VRS)$ for $k = 1, \cdots, K$ from the first K

constraints. The constraints (i) restrict the value of $IF_i((r/R)^k, x^k \mid VRS)$ for individual k to be

less than or equal to unity (and therefore $\ln(IF_i((r/R)^k, x^k \mid VRS) \leqq \ln 1$), as required by the

definition of $IF_i((r/R)^k, x^k \mid VRS)$. The restrictions (iii) impose symmetry and (ii) imposes

homogeneity of degree $+1$ in inputs.

We note that the problem in (6.4.1) imposes no restrictions on the scale properties of the

technology, thus the associated technical efficiency measures are relative to the *VRS* or *Variable*

Returns Technology. To calculate the Farrell type decomposition we need to calculate

$IF_i((r/R)^k, x^k \mid CRS)$, i.e., technical efficiency relative to technology satisfying constant returns.

This is achieved by calculating (6.4.1) with the following additional set of restrictions

(6.4.1′) $\sum_{m=1}^{M}\left(\alpha_m + \sum_{m'=1}^{M} \alpha_{mm'}(\ln(r_{km}/R_k)) + \sum_{n=1}^{N} \gamma_{mn} \ln x_{kn}\right) = 1, k = 1, \cdots, K.$

The restrictions in (6.4.1′) are derived from requiring the revenue indirect scale elasticity to equal unity, i.e., $\varepsilon_{ID_i} = (\sum_{m=1}^{M} \frac{\partial ID_i(r/R,x)}{\partial(r_m/R)} \frac{(r_m/R)}{ID_i(r/R,x)})^{-1} = 1$, see (5.3.13). $IF_i((r/R)^k, x^k \mid CRS)$ is then calculated from the set of constraints (i) like those in (6.4.1).

We can use these two measures of revenue indirect technical efficiency to calculate scale efficiency using the definition in (6.3.4), i.e., $IS_i((r/R)^k, x^k) = IF_i((r/R)^k, x^k \mid CRS)/ IF_i((r/R)^k, x^k \mid VRS)$. Returns to scale can be determined by calculating ε_{ID_i} when (6.4.1) is calculated without (6.4.1′).

In order to derive $IO_i((r/R)^k, p^k, x^k)$ we need to calculate a parametric analog of $IC((r/R)^k, p^k)$. Thus we calculate the industry frontier analog of the translog indirect cost function (see (4.2.22)) satisfying constant returns as

(6.4.2)
$$\min_{(\alpha,\beta,\gamma)} \sum_{k=1}^{K} \Bigg[\alpha_o + \sum_{m=1}^{M} \alpha_m \ln(r_{km}/R_k) + \sum_{n=1}^{N} \beta_n \ln p_{kn}$$
$$+ \frac{1}{2} \sum_{m=1}^{M} \sum_{m'=1}^{M} \alpha_{mm'}(\ln(r_{km}/R_k))(\ln(r_{km'}/R_k))$$
$$+ \frac{1}{2} \sum_{n=1}^{N} \sum_{n'=1}^{N} \beta_{nn'}(\ln p_{kn})(\ln p_{kn'}) + \sum_{m=1}^{M} \sum_{n=1}^{N} \gamma_{mn}(\ln(r_{km}/R_k))(\ln p_{kn}) \Bigg]$$

s.t. (i) $\alpha_o + \sum_{m=1}^{M} \alpha_m \ln(r_{km}/R_k) + \sum_{n=1}^{N} \beta_n \ln p_{kn}$
$$+ \frac{1}{2} \sum_{m=1}^{M} \sum_{m'=1}^{M} \alpha_{mm'}(\ln(r_{km}/R_k))(\ln(r_{km'}/R_k))$$
$$+ \frac{1}{2} \sum_{n=1}^{N} \sum_{n'=1}^{N} \beta_{nn'}(\ln p_{kn})(\ln p_{kn'}),$$
$$\sum_{m=1}^{M} \sum_{n=1}^{N} \gamma_{mn}(\ln(r_{km}/R_k))(\ln p_{kn}) \leqq \ln C_k, k = 1, \cdots, K,$$

(ii) $\sum_{m=1}^{M} \left[\alpha_m + \sum_{m'=1}^{M} \alpha_{mm'}(\ln(r_{km}/R_k)) + \sum_{n=1}^{N} \gamma_{mn}(\ln p_{kn}) \right] = 1,$
$k = 1, \cdots, K,$

(iii) $\sum_{n=1}^{N} \beta_n = 1,$

$\sum_{n=1}^{N} \beta_{nn'} = \sum_{m=1}^{M} \gamma_{mn} = 0, m = 1, \cdots, M, n = 1, \cdots, N,$

(iv) $\alpha_{mm'} = \alpha_{m'm}, m = 1, \cdots, M, m' = 1, \cdots, M,$

127

$$\beta_{nn'} = \beta_{n'n}, n = 1, \cdots, N, n' = 1, \cdots, N.$$

The constraints (i) - (iv) impose restrictions similar to those in (6.4.1). Note that (ii) imposes CRS. Individual measures of $IO_i((r/R)^k, p^k, x^k)$ can be derived by calculating the relevant value of the first set of constraints, $k = 1, \cdots, K$ (which yields $IC((r/R)^k, p^k)$ and dividing it by $C_k, k = 1, \cdots, K$. Finally, allocative efficiency can be calculated for each observation $k = 1, \cdots, K$ by exploiting the definition $IA_i((r/R)^k, p^k, x^k) = IO_i((r/R)^k, p^k, x^k)/IF_i((r/R)^k, x^k)$.

We turn next to a parameterized version of the decomposition of cost indirect revenue efficiency. We begin with the calculation of the cost indirect output measure of technical efficiency. Recall from Section 6.2 that this measure is the reciprocal of the cost indirect output distance function which was parameterized as a translog function in Chapter 5, see (5.2.18). Analogous to that specification we can write the following programming problem.

$$(6.4.3) \quad \min_{(\alpha,\beta,\gamma)} \sum_{k=1}^{K} \left[\alpha_o + \sum_{n=1}^{N} \alpha_n \ln(p_{kn}/C_k) + \sum_{m=1}^{M} \beta_n \ln u_{km} \right.$$
$$+ \frac{1}{2} \sum_{n=1}^{N} \sum_{n'=1}^{N} \alpha_{nn'} (\ln(p_{kn}/C_k))(\ln(p_{kn'}/C_k))$$
$$+ \frac{1}{2} \sum_{m=1}^{M} \sum_{m'=1}^{M} \beta_{mm'} (\ln u_{km})(\ln u_{km'})$$
$$\left. + \sum_{n=1}^{N} \sum_{m=1}^{M} \gamma_{nm} (\ln(p_{kn}/C_k))(\ln u_{km}) \right]$$

s.t. (i) $\alpha_o + \sum_{n=1}^{N} \alpha_n (\ln(p_{kn}/C_k)) + \sum_{m=1}^{M} \beta_m \ln u_{km}$
$$+ \frac{1}{2} \sum_{n=1}^{N} \sum_{n'=1}^{N} \alpha_{nn'} (\ln(p_{kn}/C_k))(\ln(p_{kn'}/C_k))$$
$$+ \frac{1}{2} \sum_{m=1}^{M} \sum_{m'=1}^{M} \beta_{mm'} (\ln u_{km})(\ln u_{km'})$$
$$+ \sum_{n=1}^{N} \sum_{m=1}^{M} \gamma_{nm} (\ln(p_{kn}/C_k))(\ln u_{km}) \geqq 0, k = 1, \cdots, K,$$

(ii) $\sum_{m=1}^{M} \beta_m = 1,$

$$\sum_{m=1}^{M} \beta_{mm'} = \sum_{m=1}^{M} \gamma_{nm} = 0, m = 1, \cdots, M, n = 1, \cdots, N,$$

(iii) $\alpha_{nn'} = \alpha_{n'n}, n = 1, \cdots, N, n' = 1, \cdots, N,$

$$\beta_{mm'} = \beta_{m'm}, m = 1, \cdots, M, m' = 1, \cdots, M.$$

Further, if one wishes to have a decomposition analogous to the original Farrell decomposition, the technology should satisfy constant returns to scale. This implies that an additional K restrictions are required:

(6.4.3′) $\sum_{n=1}^{N}[\alpha_n + \sum_{n'=1}^{N} \alpha_{nn'}(\ln(p_{kn'}/C_k)) + \sum_{m=1}^{M} \gamma_{nm} \ln u_{km}] = 1, k = 1, \cdots, K.$

This additional restriction is due to setting the cost indirect output based measure of scale elasticity equal to unity, i.e.,

$$\varepsilon_{ID_o} = \sum_{n=1}^{N} \frac{\partial ID_o(p/C, u)(p_n/C)}{\partial(p_n/C)ID_o(p/C, u)} = 1,$$

for all k (see (5.3.10)).

Thus the $k = 1, \cdots, K$ individual cost indirect output measures of technical efficiency satisfying constant returns, $IF_o((p/C)^k, u^k \mid CRS)$, can be calculated from the (i) constraints in (6.4.2). If the restriction (6.4.2′) is not imposed, the resulting measures are relative to a variable returns to scale technology, i.e., the first K constraints yield $IF_o((p/C)^k, u^k \mid VRS)$. From these two measures, using the definition in (6.3.10) we can calculate cost indirect scale efficiency as

(6.4.4) $IS_o((p/C)^k, u^k) = IF_o((p/C)^k, u^k \mid CRS)/IF_o((p/C)^k, u^k \mid VRS), k = 1, \cdots, K.$

Turning to "overall" cost indirect efficiency, i.e., $IO_o((p/C)^k, r^k, u^k)$, we need to calculate cost indirect maximal revenue relative to a constant returns to scale technology. We then have the industry version of the problem from (4.1.29) with the additional constraint that

$\varepsilon_{IR} = -\sum_{n=1}^{N} \frac{\partial IR(p_{kn}/C_k, r_k)(p_{kn}/C_k)}{\partial(p_{kn}/C_k)IR(p_{kn}/C_k, r_k)} = 1, k = 1, \cdots, K$ (see (5.3.14) and constraint (ii) below) to ensure constant returns to scale, i.e.,

(6.4.5) $\displaystyle\max_{(\alpha,\beta,\gamma)} \sum_{k=1}^{K} \Bigg[\alpha_o + \sum_{n=1}^{N} \alpha_n \ln(p_{kn}/C_k) + \sum_{m=1}^{M} \beta_m \ln r_{km}$

$\displaystyle + \frac{1}{2} \sum_{n=1}^{N} \sum_{n'=1}^{N} \alpha_{nn'}(\ln(p_{kn}/C_k))(\ln(p_{kn'}/C_k))$

$\displaystyle + \frac{1}{2} \sum_{m=1}^{M} \sum_{m'=1}^{M} \beta_{mm'}(\ln r_{km})(\ln r_{km'})$

$\displaystyle + \sum_{n=1}^{N} \sum_{m=1}^{M} \gamma_{nm}(\ln(p_{kn}/C_k))(\ln r_{km}) \Bigg]$

s.t.   (i) $\displaystyle \alpha_o + \sum_{n=1}^{N} \alpha_n(\ln(p_{kn}/C_k)) + \sum_{m=1}^{M} \beta_m \ln r_{km}$

129

$$+\frac{1}{2}\sum_{n=1}^{N}\sum_{n'=1}^{N}\alpha_{nn'}(\ln(p_{kn}/C_k))(\ln(p_{kn'}/C_k))$$

$$+\frac{1}{2}\sum_{m=1}^{M}\sum_{m'=1}^{M}\beta_{mm'}(\ln r_{km})(\ln r_{km'})$$

$$+\sum_{n=1}^{N}\sum_{m=1}^{M}\gamma_{nm}(\ln(p_{kn}/C_k))(\ln r_{km}) \geqq \ln R_k, k = 1, \cdots, K,$$

(ii) $\displaystyle\sum_{n=1}^{N}\left[\alpha_n + \sum_{n'=1}^{N}\alpha_{nn'}(\ln(p_{kn}/C_k) + \sum_{m=1}^{M}\ln r_{km}\right] = 1,$

$k = 1, \cdots, K,$

(iii) $\displaystyle\sum_{m=1}^{M}\beta_m = 1, \sum_{m'=1}^{M}\beta_{mm'} = \sum_{m=1}^{M}\gamma_{nm} = 0, m = 1, \cdots, M,$

$n = 1, \cdots, N,$

(iv) $\alpha_{nn'} = \alpha_{n'n}, \beta_{mm'} = \beta_{m'm}, n, n' = 1, \cdots, N, m, m' = 1, \cdots, M.$

The first (i) constraints yield $IR((p/C)^k, r^k), k = 1, \cdots, K$. Dividing by $R_i, k = 1, \cdots, K$ yields $IO_o((p/C)^k, r^k, u^k)$.

Finally, $IA_o((p/C)^k, r^k, u^k)$ can be calculated as

(6.4.6) $\qquad IA_o((p/C)^k, r^k, u^k) = IO_o((p/C)^k, r^k, u^k)/IF_o((p/C)^k, r^k, u^k), k = 1, \cdots, K.$

6.P Problems

(6.P.1) Define direct input and output measures of technical efficiency and show that they are reciprocal to the input and output distance functions, respectively.

(6.P.2) Show that the revenue indirect input measure of technical efficiency is homogeneous of degree -1 in inputs.

(6.P.3) What assumptions on $IL(r/R)$ guarantee the existence of a minimum for $IF((r/R)^k, x^k)$?

(6.P.4) Show that $IF_i((r/R)^k, x^k)$ satisfies properties like $(ID_i.1 - ID_i.6)$.

(6.P.5) Show that $IF_o((p/C)^k, u^k)$ satisfies properties like $(ID_o.1 - ID_o.6)$.

(6.P.6) Show that $IC(r/R, p) = \inf_x\{px : ID_i(r/R, x) \leqq 1\}$.

(6.P.7) Prove proposition (6.2.9).

(6.P.8)   Derive the homogeneity properties of $IO_i((r/R)^k, p^k, x^k)$ and $IO_o((p/C)^k, r^k, u^k)$.

(6.P.9)   Derive the direct Farrell input based decomposition by substituting u^k for $(r/R)^k$ in the revenue indirect cost function and technical efficiency measures.

(6.P.10)   Derive direct Farrell measures and decomposition for the output based case. That is, substitute x^k for $(p/C)^k$ in the cost indirect revenue and technical efficiency functions.

(6.P.11)   Show $IF_o((p/C)^c, u^c \mid VRS)$ and $IS_o((p/C)^c, u^c)$ in Figure 6.4.

Notes

This chapter is based on Farrell (1957), Färe, Grosskopf and Lovell (1985), (1988), (1990). The parametric programming models in Section 6.4 are similar in spirit to the industry frontier model due to Aigner and Chu (1968), and are also related to similar industry frontier models due to Førsund and Hjalmarsson (1987). For a discussion of alternative approaches to calculating frontiers and efficiency, including stochastic frontiers, see Lovell and Schmidt (1988).

Chapter 7

Productivity

7.0 Introduction

In the previous chapter we focused on the measurement of efficiency, where efficiency of a given activity was gauged relative to frontier performance. It was shown that distance functions and their dual representations are a natural way of modelling various frontiers and of calculating the deviation of individual activities from those frontiers.

In this chapter we turn our attention to the measurement of productivity. Here we define productivity as the product of changes in performance (efficiency) and changes in the frontier of technology over time. At a disaggregated level we are tracking an activity or firm's performance over time; at a more aggregated level we are trying to measure shifts in industry frontier technology. Clearly, efficiency and productivity are related: one of the goals of this chapter is to build productivity indexes that explicitly recognize that relationship.

In addition to explictly including efficiency in the calculation of productivity, and in keeping with the overall theme of this book, we pay particular attention to measuring productivity of firms which face budget or revenue constraints. This allows us to generalize existing productivity indexes, thus providing a broader menu of models, and through dual representations, a broader menu of data choices to those wishing to measure productivity.

We begin with the family of productivity indexes associated with the name Malmquist. These indexes were originally defined by Caves, Christensen and Diewert (1982) as ratios of distance functions. Here we define the productivity indexas as geometric means of ratios of distance functions calculated at adjacent time periods, i.e., productivity is calculated as the geometric mean of the shift in the frontier evaluated at the base and the comparison period. We introduce an indirect version of the Malmquist output based productivity index, in which firms face a budget constraint. This yields indexes which may be calculated based on cost indirect output distance functions. Noting that these distance functions are the reciprocal of Farrell type indirect efficiency measures, we provide a decomposition of this indirect productivity measure into changes in efficiency and changes in frontier technology. Having shown in Section 5.1 when indirect and direct output distance functions coincide, we can also show when indirect and direct Malmquist output productivity measures will coincide. A parallel development follows based on direct and indirect input based indexes.

Since Malmquist productivity indexes are rarely calculated in practice, we devote two sections of this chapter to the problem of calculating these indexes. We provide two approaches, both of which employ linear programming methods, one nonparametric, the other parametric. In the nonparametric approach we exploit the relationship between Farrell type efficiency measures and distance functions, i.e., we show how to calculate the component distance functions as the reciprocals of efficiency measures which are calculated relative to a piecewise linear representation of technology. As an alternative, we also introduce a series of parameterized programming problems to calculate these component distance functions using a translog specification based on the frontier approach suggested by Aigner and Chu.

In Section 7.4 we turn to another (nonparametric) index of productivity – the Fisher ideal index, which is the geometric mean of the corresponding Laspeyres and Paasche quantity indexes. Thus, in contrast to the original Malmquist indexes, the Fisher index is constructed in value terms. We show however, that in fact these two indexes coincide under certain assumptions. This relationship is developed for both input and output quantity indexes. Having established these relationships, we turn to the indirect Malmquist indexes and their relationship to Fisher ideal indexes.

The productivity index which is most widely used in recent applied work is probably the Törnqvist index. Its popularity lies in its ease of computation (it is nonparametric) and the fact that it is "exact" if underlying technology is translog. In Section 7.5 we demonstrate the relationship between the Törnqvist and Malmquist indexes of productivity.

7.1 Malmquist Productivity Indexes

In this section we introduce various productivity indexes defined in terms of distance functions. The distance function foundation of these indexes prompts us to call them Malmquist indexes.

Recall that the cost indirect output distance function is defined as

$$(7.1.1) \qquad ID_o(p/C, u) = \inf\{\theta > 0 : (u/\theta) \in IP(p/C)\},$$

where $IP(p/C) = \{u : u \in P(x), px \leqq C\}$. Moreover, whenever (p/C) is strictly positive, $IP(p/C) = \{u : C(u,p) \leqq C\}$, where $C(u,p)$ is the cost function.

For two distinct firms or periods "o" and "1" we define the *Malmquist Cost Indirect Output Based Productivity Index* as

$$(7.1.2) \qquad IM_o^1((p/C)^1, u^1, (p/C)^0, u^0) = \left[\frac{ID_o^0((p/C)^1, u^1)}{ID_o^0((p/C)^0, u^0)} \frac{ID_o^1((p/C)^1, u^1)}{ID_o^1((p/C)^0, u^0)}\right]^{\frac{1}{2}}$$

or equivalently as

$$(7.1.3) \qquad IM_o^1((p/C)^1, u^1, (p/C)^0, u^0)) = \frac{ID_o^1((p/C)^1, u^1)}{ID_o^0((p/C)^0, u^0)} \left[\frac{ID_o^0((p/C)^1, u^1)}{ID_o^1((p/C)^1, u^1)} \frac{ID_o^0((p/C)^0, u^0)}{ID_o^1((p/C)^0, u^0)}\right]^{\frac{1}{2}} .$$

The ratio outside the bracket in expression (7.1.3) measures the change in or ratio of technical efficiency between "o" and "1". Thus if both observations are technically efficient, this ratio is equal to one, and in addition the index becomes $(ID_o^0((p/C)^1, u^1)/ID_o^1((p/C)^0, u^0))^{\frac{1}{2}}$. The ratios inside the bracket measure the change in the frontier. Figure 7.1 where $(p/C)^o = (p/C)'$ illustrates.

The output vector u^0 is feasible under $IP^0((p/C)^0)$ and u^1 is feasible under $IP^1((p/C)^1)$. The quotient outside the bracket equals $(0d/0e)/(0a/0b)$. The expression inside the bracket is $[(0d/0f)/(0d/0e)][(0a/0b)/(0a/0c)]$, thus (7.1.3) may be expressed in terms of the above distances as

$$(7.1.4) \qquad \left(\frac{0d}{0e} \frac{0b}{0a}\right) \left[\frac{0e}{0f} \frac{0c}{0b}\right]^{\frac{1}{2}},$$

and the productivity index measures change in efficiency times the geometric mean of the change in the frontier evaluated along the ray through u^0 $(0c/0b)$ and along the ray through u^1 $(0e/0f)$.

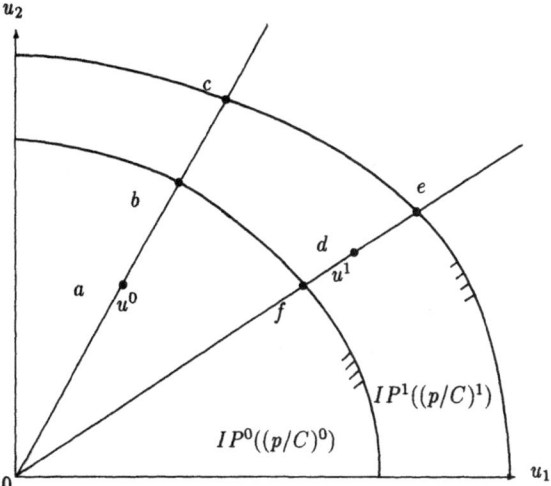

Figure 7.1: The Malmquist Cost Indirect Productivity Index

The index in (7.1.3) may take on any nonnegative value. A value of unity would imply that the combined efficiency change and frontier change resulted in no net improvement or detioration. More specifically:

$$IM_o^1((p/C)^1, u^1, (p/C)^0, u^0) > 1 \text{ implies improvement from "0" to "1"}$$

$$IM_o^1((p/C)^1, u^1, (p/C)^0, u^0) < 1 \text{ implies deterioration from "0" to "1".}$$

To obtain geometric mean of the Malmquist productivity indexs introduced by Caves, Christensen and Diewert (1982), one merely substitutes direct output distance functions for the indirect distance functions included in (7.1.2). Thus the *Malmquist Output Based Productivity Index* may be defined as

$$(7.1.5) \qquad M_o^1(x^1, u^1, x^0, u^0) = \left[\frac{D_o^0(x^1, u^1)}{D_o^0(x^0, u^0} \frac{D_o^1(x^1, u^1)}{D_o^1(x^0, u^0)} \right]^{\frac{1}{2}}.$$

Of course, we can also decompose the direct measure $M_0^1(x^1, u^1, x^0, u^0)$ into a product of the two component measures, change in technical efficiency and the geometric mean of the change in the frontier. Again, improvements over time are indicated by values of (7.1.5) greater than unity.

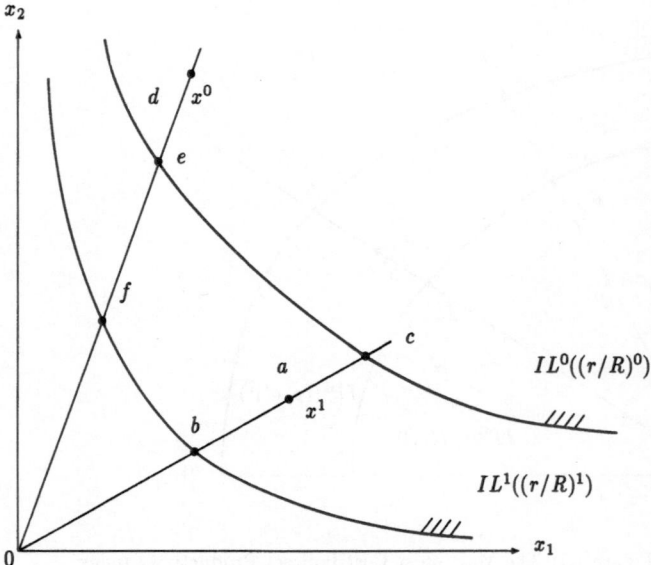

Figure 7.2: The Malmquist Revenue Indirect Productivity Index

Turning to the input based productivity measures, recall the definition of indirect input distance function,

(7.1.6) $ID_i(r/R, x) = \sup\{\lambda > 0 : (x/\lambda) \in IL(r/R)\}$,

where $IL(r/R) = \{x : x \in L(u), ru \geqq R\}$. In terms of the revenue indirect input distance function the *Malmquist Revenue Indirect Input Based Productivity Index* is defined as

(7.1.7) $IM_i^1((r/R)^1, x^1, (r/R)^0, x^0) = \left[\dfrac{ID_i^0((r/R)^1, x^1)}{ID_i^0((r/R)^0, x^0)} \dfrac{ID_i^1((r/R)^1, x^1)}{ID_i^1((r/R)^0, x^0)} \right]^{\frac{1}{2}}$,

or equivalently as

(7.1.8) $IM_i^1((r/R)^1, x^1, (r/R)^0, x^0) = \dfrac{ID_i^1((r/R)^1, x^1)}{ID_i^0((r/R)^0, x^0)} \left[\dfrac{ID_i^0((r/R)^1, x^1)}{ID_i^1((r/R)^1, x^1)} \dfrac{ID_i^0((r/R)^0, x^0)}{ID_i^1((r/R)^0, x^0)} \right]^{\frac{1}{2}}$,

where the ratio in front of the bracket in (7.1.8) measures the change in technical efficiency between "o" and "1". If the observations are efficient relative to their own reference technology, then (7.1.8) becomes $(ID_i^0((r/R)^1, x^1)/ID_i^1((r/R)^0, x^0))^{\frac{1}{2}}$. The ratios inside the bracket in expression (7.1.8) measure the shift in the frontier. Figure 7.2 where $(r/R)^0 = (r/R)^1$ illustrates.

136

In the figure, $x^0 \in IL^0((r/R)^0)$ and $x^1 \in IL^1((r/R)^1)$, where the "0" input set is contained in the "1" input set. The measure of efficiency change is given by $(0a/0b)/(0d/0e))$, and the change in the frontier is expressed by $[(0d/0e)/(0d/0f)][(0a/0c)/(0a/0b)]$. Thus in terms of the distances in Figure 7.2, the productivity index (7.1.8) becomes

$$(7.1.9) \qquad \left(\frac{0e}{0d} \frac{0a}{0b} \right) \left[\frac{0f}{0e} \frac{0b}{0c} \right]^{\frac{1}{2}},$$

and our index is the product of the change in efficiency $(\frac{0e}{0d} \frac{0a}{0b})$ and the geometric mean of the shift in frontier evaluated along the ray through x^0 $(0f/0e)$ and evaluated along the ray through x^2 $(0b/0c)$. $IM_i^1((r/R)^1, x^1, (r/R)^0, x^0) < 1$ denotes improvement between "0" and "1".

Following Caves, Christensen and Diewert, we may also define an input based productivity index constructed as ratios of direct rather than indirect input distance functions. Specifically, the *Malmquist Input Based Productivity Index* is

$$(7.1.10) \qquad M_i^1(u^1, x^1, u^0, x^0) = \left[\frac{D_i^0(u^1, x^1)}{D_i^0(u^0, x^0)} \frac{D_i^1(u^1, x^1)}{D_i^1(u^0, x^0)} \right]^{\frac{1}{2}}.$$

As before, this is actually the geometric mean of two indexas as defined by Caves, Christensen and Diewert.

We leave to the reader the formulation of the decomposition of (7.1.10), and note that like the indirect Malmquist input based productivity index, improvements between "0" and "1" yield values of the index of less than unity.

7.2   Piecewise Linear Calculation of Productivity

In Section 3.3, the direct and indirect technologies are modeled as piecewise linear. These models are used in this section as reference technologies, relative to which efficiency and productivity are measured. Suppose that for each $t = 0, 1$, there are $k = 1, \cdots, K^t$ observations on inputs $x^{k,t} = (x_{k1}^t, \cdots, x_{kN}^t)$, outputs $u^{k,t} = (u_{k1}^t, \cdots, u_{kM}^t)$, input prices $p^{k,t} = (p_{k1}^t, \cdots, p_{kN}^t)$ and cost $C^{k,t}$. We assume that inputs and outputs may include some zeros as specified in condition (3.3.1) and that $p^{k,t} \in \Re_+^N$, and $p^{k,t} \neq 0$, moreover $C^{k,t} > 0, k = 1, \cdots, K^t, t = 0, 1$.

Recall that distance functions are the reciprocals of the corresponding Farrell measures of technical efficiency. For example, $(ID_o((p/C)^{k,t}, u^{k,t}))^{-1} = IF_o((p/C)^{k,t}, u^{k,t})$, thus some of the calculations in this section are replicated in Chapter 6. Here we suppose that the parent

137

technology exhibits constant returns to scale, and that both inputs and outputs are strongly disposable, i.e., its graph is modeled by (3.3.2). Obviously, more general parent technologies may be chosen, see Section 3.3 for the associated restrictions required in the piecewise linear model. The reciprocal formulation is employed in order to maintain linearity.

Before proceeding to the calculations, we point out several difficulties which one may encounter in implementing this approach. The first difficulty arises due to the fact that we will be calculating what we refer to as "mixed period" distance functions. For example, $D_o^0(x^1, u^1)$, where period "1" data is evaluated relative to period "0" technology. In contrast to the own period distance functions such as $D_o^0(x^0, u^0)$, the observation being evaluated may not be an element of the reference technology, and the possibility of nonexistence of positive solutions arises. Some, but not all, of these problems may be avoided by assuming constant returns to scale. This may be seen in Figure 7.3 in which we consider a scalar input, scalar output case. Technology in period "0" satisfies variable returns to scale. Observation $(k, 1)$ is not an element of period "0" technology, i.e., $(k, 1) \notin GR^0$. If one tries to expand $(k, 1)$ in the output direction as much as possible (i.e., solve $D_o^0(x^{k,1}, u^{k,1})$), there is no solution. Note that a solution does exist in this case if the technology is extended to satisfy $NIRS$ or CRS. Disappearing goods may also cause nonexistence of positive solutions in the mixed period case even under CRS, see problem (7.P.3). Finally if data is missing for an observation in some period, then that observation cannot be given a productivity measure of that period.

In order to calculate the Malmquist indirect output based productivity index (7.1.2) or (7.1.3), four different linear programs must be solved for each observation k. The first is

$$
(7.2.1) \quad (ID_o^0((p/C)^{k,0}, u^{k,0}))^{-1} = \max_{\theta, z, x} \theta
$$

$$
s.t. \quad \sum_{k=1}^{K^0} z_k u_{km}^0 \geq \theta u_{km}^0, m = 1, \cdots, M,
$$

$$
\sum_{k=1}^{K^0} z_k x_{kn}^0 \leq x_n, n = 1, \cdots, N,
$$

$$
\sum_{n=1}^{N} p_{kn}^0 x_n \leq C_k^0,
$$

$$
z_k \geq 0, k = 1, \cdots, K^0.
$$

The first component (7.2.1) of the output based indirect Malmquist productivity index computes

138

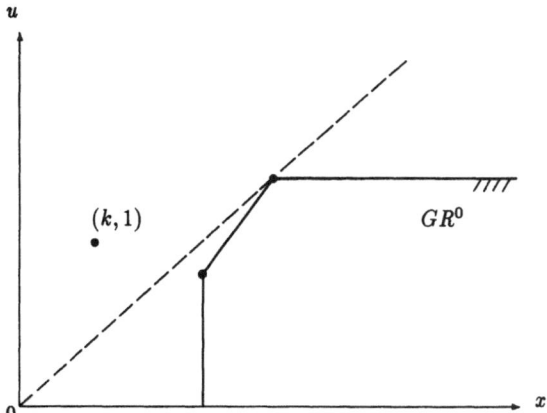

Figure 7.3: Nonexistence of $ID_o^0((p/C)^{k,1}, u^{k,1})$ with VRS technology.

the indirect output based Farrell technical efficiency score of observation k^0 relative to the reference technology of the same period. This component is one of those in the ratio "outside" the bracket in (7.1.3).

The second component is $ID_o^1((p/C)^{k,1}, u^{k,1})$, which can be calculated as the solution to the following problem.

$$(7.2.2) \quad (ID_o^1((p/C)^{k,1}, u^{k,1}))^{-1} = \max_{\theta, z, x} \theta$$

$$s.t. \quad \sum_{k=1}^{K^1} z_k u_{km}^1 \geqq \theta u_{km}^1, m = 1, \cdots, M,$$

$$\sum_{k=1}^{K^1} z_k x_{kn}^1 \leqq x_n, n = 1, \cdots, N,$$

$$\sum_{n=1}^{N} p_{kn}^1 x_n \leqq C_k^1,$$

$$z_k \geqq 0, k = 1, \cdots, K^1.$$

The reciprocal of the second component $ID_o^1((p/C)^{k,1}, u^{k,1})$ measures the indirect output based Farrell technical efficiency of observation k at $t = 1$, i.e., k^1 relative to the reference technology in the same period.

In contrast to the first two components of the productivity index, the last two components measure Farrell technical efficiency with the reference technology and the observation at different time periods. The first of these components is

$$(7.2.3) \quad \left(ID_o^0 \left((p/C)^{k,1}, u^{k,1} \right) \right)^{-1} = \max_{\theta,z,x} \theta$$

$$s.t. \quad \sum_{k=1}^{K^0} z_k u_{km}^0 \geq \theta u_{km}^1, m = 1, \cdots, M,$$

$$\sum_{k=1}^{K^0} z_k x_{kn}^0 \leq x_n, n = 1, \cdots, N,$$

$$\sum_{n=1}^{N} p_{kn}^1 x_n \leq C_k^1,$$

$$z_k \geq 0, k = 1, \cdots, K^0.$$

In this problem, observation k^1 is evaluated relative to the reference technology at $t = 0$. In Figure 7.1 this amounts to computing $(0d/0f)$. The fourth component evaluates observation k^0 relative to the reference technology at $t = 1$. In Figure 7.1 this equals $(0a/0c)$.

$$(7.2.4) \quad \left(ID_o^1 \left((p/C)^{k,0}, u^{k,0} \right) \right)^{-1} = \max_{\theta,z,x} \theta$$

$$s.t. \quad \sum_{k=1}^{K^1} z_k u_{km}^1 \geq \theta u_{km}^0, m = 1, \cdots, M,$$

$$\sum_{k=1}^{K^1} z_k x_{kn}^1 \leq x_n, n = 1, \cdots, N,$$

$$\sum_{n=1}^{N} p_{kn}^0 x_n \leq C_k^0,$$

$$z^k \geq 0, k = 1, \cdots, K^1.$$

The four linear programming problems (7.2.1) – (7.2.4) make up the component measures for the indirect Malmquist index (7.1.2) or in decomposed form (7.1.3) for observation $k^t, t = 0, 1$. These can then be calculated for each $k = 1, \cdots, K$ for every pair of adjacent periods $t = 0, 1$.

The output based direct Malmquist productivity index may also be computed by solving four different linear programming problems. Here we show how two of the components are calculated. Suppose that data on inputs and outputs are available for two periods, and that the data satisfies the same conditions as for the indirect measures. Then for observation k^0 we compute

$$(7.2.5) \quad \left(D_o^0(x^{k,0}, u^{k,0}) \right)^{-1} = \max_{\theta,z} \theta$$

$$s.t. \quad \sum_{k=1}^{K^0} z_k u_{km}^0 \geq \theta u_{km}^0, m = 1, \cdots, M,$$

$$\sum_{k=1}^{K^0} z_k x_{kn}^0 \leq x_{kn}^0, n = 1, \cdots, N,$$

$$z_k \geq 0, k = 1, \cdots, K^0.$$

Again, constant returns to scale and strong disposability are imposed on the parent technology. See Section 3.3 for how these may be relaxed. The linear programming problem (7.2.5) measures the output based Farrell technical efficiency of observation k^0 relative to the reference technology of the same period. The second component in the direct Malmquist output productivity index $D_o^1(x^1, u^1)$ also measures Farrell technical efficiency relative to the reference technology of the same period, thus we do not write it out. Instead, we consider observation k^1 in relation to the reference technology at "0". This component of $M_o^1(x^1, u^1, x^0, u^0)$ is computed as

$$(7.2.6) \quad \left(D_o^0(x^{k,1}, u^{k,1}) \right)^{-1} = \max_{\theta, z} \theta$$

$$s.t. \quad \sum_{k=1}^{K^0} z_k u_{km}^0 \geq \theta u_{km}^1, m = 1, \cdots, M,$$

$$\sum_{k=1}^{K^0} z_k x_{kn}^0 \leq x_{kn}^1, n = 1, \cdots, N,$$

$$z_k \geq 0, k = 1, \cdots, K^0.$$

The input and output constraints have the two periods "0" and "1" on opposite sides of the inequalities, indicating that the observation, here k^1, is compared to the reference technology of a different period, here "0". The fourth and last component in $M_o^1(x^1, u^1, x^0, u^0)$, is similar in character to the third component, and it can be computed as in (7.2.6) by interchanging "1" and "0", which we leave to the reader.

Turning to the indirect input based Malmquist productivity index, we note that both formulations (7.1.7) and (7.1.8) consist of four components. Here we give their linear programming formulations, under the assumption that the parent technology exhibits constant returns to scale with strongly disposable inputs and outputs, see (3.3.1). Additional assumptions may be imposed; such alternatives are discussed in Section 3.3. We suppose that for each $t = 0, 1$, there are $k = 1, \cdots, K^t$ observations on inputs $x^{k,t} = (x_{k1}^t, \cdots, x_{kN}^t)$, outputs $u^{k,t} = (u_{k1}^t, \cdots, u_{kM}^t)$, output prices $r^{k,t} = (r_{k1}^t, \cdots, r_{kM}^t)$ and target revenues $R^{k,t}$. The inputs and outputs are assumed to satisfy (3.3.1) at each t, and we assume that $r^{t,1} \neq 0$ and $R^{k,t} > 0, k = 1, \cdots, K^t, t = 0, 1$. Again, recall that distance functions are reciprocal to the

141

corresponding Farrell measures of technical efficiency, so that we can make use of some of the programming problems from Chapter 6. The first component of the indirect input based productivity measure computes the "efficiency" of observation k^0 relative to the frontier of the same period. In terms of Figure 7.2 it measures $(0d/0e)$, and in general may be calculated as

$$
(7.2.7) \quad \left(ID_i^0\left((r/R)^{k,0}, x^{k,0}\right)\right)^{-1} = \min_{\lambda, z, u} \lambda
$$

$$
s.t. \quad \sum_{k=1}^{K^0} z_k u_{km}^0 \geqq u_m^0, m = 1, \cdots, M,
$$

$$
\sum_{k=1}^{K^0} z_k x_{kn}^0 \leqq \lambda x_{kn}^0, n = 1, \cdots, N,
$$

$$
\sum_{m=1}^{M} r_{km}^0 u_m \geqq R_k^0,
$$

$$
z_k \geqq 0, k = 1, \cdots, K^0.
$$

Next we turn to the second component, which measures the technical efficiency of k^1 relative to the frontier at the same period, and which may be computed as

$$
(7.2.8) \quad \left(ID_i^1\left((r/R)^{k,1}, x^{k,1}\right)\right)^{-1} = \min_{\lambda, z, u} \lambda
$$

$$
s.t. \quad \sum_{k=1}^{K^1} z_k u_{km}^1 \geqq u_m, m = 1, \cdots, M,
$$

$$
\sum_{k=1}^{K^1} z_k x_{kn}^1 \leqq \lambda x_{kn}^1, n = 1, \cdots, N,
$$

$$
\sum_{m=1}^{M} r_{km}^1 u_m \geqq R_k^1,
$$

$$
z_k \geqq 0, k = 1, \cdots, K^1.
$$

In terms of Figure 7.2, (7.2.8) computes the ratio $(0a/0b)$, which is the indirect technical efficiency measure of k^1 relative to the frontier at the same period "1". The third component involves mixed periods, i.e., observation k^0 is related to the frontier at "1".

$$
(7.2.9) \quad \left(ID_i^1\left((r/R)^{k,0}, x^{k,0}\right)\right)^{-1} = \min_{\lambda, z, u} \lambda
$$

$$
s.t. \quad \sum_{k=1}^{K^1} z_k u_{km}^1 \geqq u_m, m = 1, \cdots, M,
$$

$$
\sum_{k=1}^{K^1} z_k x_{kn}^1 \leqq \lambda x_{kn}^0, n = 1, \cdots, N,
$$

142

$$\sum_{m=1}^{M} r_{km}^0 u_m \geqq R_k^0,$$

$$z_k \geqq 0, k = 1, \cdots, K^1.$$

Note that in (7.2.9) both period notations "0" and "1" occur, since observation k^0 is compared to the reference technology at "1". Although it is quite obvious, one should mention that in the output constraints, the $u_m, m = 1, \cdots, M$ are variables, which is why they do not contain a k or "0" or "1". In Figure 7.2, problem (7.2.9) measures $(0d/0f)$.

The last component of the indirect input based productivity index, like (7.2.9), involves mixed periods. Here, the observation k^1 is compared to the reference technology of period "0". In particular,

$$(7.2.10) \quad \left(ID_i^0 \left((r/R)^{k,1}, x^{k,1} \right) \right)^{-1} = \min_{\lambda,z,u} \lambda$$

$$s.t. \quad \sum_{k=1}^{K^0} z_k u_{km}^1 \geqq u_m, m = 1, \cdots, M,$$

$$\sum_{k=1}^{K^0} z_k x_{kn}^0 \leqq \lambda x_{kn}^1, n = 1, \cdots, N,$$

$$\sum_{m=1}^{M} r_{km}^1 u_m \geqq R_k^1,$$

$$z_k \geqq 0, k = 1, \cdots, K^0.$$

As a reminder, the parent technology described in the constraints of (7.2.10) satisfies constant returns to scale and inputs and outputs are strongly disposable. The solutions to the four problems (7.2.7) – (7.2.10) are used to form the indirect input based productivity index for each observation $k = 1, \cdots, K$.

The direct input based Malmquist productivity index $M_i^1(u^1, x^1, u^0, x^0)$ may also be calculated by solving four linear programming problems. Here we only introduce two and leave to the reader the formulation of the remaining two. Suppose that data on inputs and outputs are available and that the data meet the same requirements as the indirect measure. Then for observation k^0 we compute

$$(7.2.11) \quad \left(D_i^0 (u^{k,0}, x^{k,0}) \right)^{-1} = \min_{\lambda,z} \lambda$$

143

$$s.t. \quad \sum_{k=1}^{K^0} z_k u_{km}^0 \geq u_{km}^0, m = 1, \cdots, M,$$

$$\sum_{k=1}^{K^0} z_k x_{kn}^0 \leq \lambda x_{kn}^0, n = 1, \cdots, N,$$

$$z_k \geq 0, k = 1, \cdots, K^0.$$

The parent technology is assumed to satisfy constant returns to scale, and inputs and outputs are strongly disposable. The input based Farrell technical efficiency is computed by (7.2.11) for k^0. Under the same assumption on technology the second component of $M_i^1(u^1, x^1, u^0, x^0)$ measures k^1 performance relative to the reference technology of the same period. This problem is left to the reader to construct.

The first of the two problems that contain mixed periods is written as

$$(7.2.12) \quad \left(D_i^0(u^{k,1}, x^{k,1}) \right)^{-1} = \min_{\lambda, z} \lambda$$

$$s.t. \quad \sum_{k=1}^{K^0} z_k u_{km}^0 \geq u_{km}^1, m = 1, \cdots, M,$$

$$\sum_{k=1}^{K^0} z_k x_{kn}^0 \leq \lambda x_{kn}^1, n = 1, \cdots, N,$$

$$z_k \geq 0, k = 1, \cdots, K^0.$$

This problem computes the efficiency of observation k in period "1" relative to the reference technology constructed from information at period "0". The fourth component of $M_i^1(u^1, x^1, u^0, x^0)$, like (7.2.12), contains mixed periods. The reader should write out the corresponding linear programming problem.

7.3 Parametric Calculation of Productivity

Parametric specifications of the various direct and indirect distance functions were introduced in Chapter 3. In this section some of these are used in the calculation of productivity. In contrast to Section 7.2 only two distance functions need to be computed in order to calculate the relevant productivity index.

Suppose that for each period $t = 0, 1$, there are $k = 1, \cdots, K^t$ observations on cost deflated input prices $(p/C)^{k,t} = (p_{k1}^t/C_k^t, \cdots, p_{kN}^t/C_k^t)$, where each $(p_{kn}^t/C_k^t) > 0$, and outputs $u^{k,t} = (u_{k1}^t, \cdots, u_{kM}^t), u_{km}^t > 0$, for all t, k and m, since our specification requires that we take logs

144

of the data. Following the ideas of Aigner and Chu (1968), the parameters of the cost indirect translog output distance function at "o" may be computed as the solution to the linear programming problem

(7.3.1) $\displaystyle\max_{(\alpha,\beta,\gamma)} \quad \sum_{k=1}^{K^0} \Bigg[\alpha_0 + \sum_{n=1}^{N} \alpha_n \ln(p_{kn}^0/C_k^0) + \sum_{m=1}^{M} \beta_m \ln u_{km}^0$

$$+ \frac{1}{2}\sum_{n=1}^{N}\sum_{n'=1}^{N} \alpha_{nn'} \left(\ln(p_{kn}^0/C_k^0) \right) \left(\ln(p_{kn'}^0/C_k^0) \right)$$

$$+ \frac{1}{2}\sum_{m=1}^{M}\sum_{m'=1}^{M} \beta_{mm'} (\ln u_{km}^0)(\ln u_{km'}^0)$$

$$+ \sum_{n=1}^{N}\sum_{m=1}^{M} \gamma_{nm} \left(\ln(p_{kn}^0/C_k^0) \right) (\ln u_{km}^0) \Bigg]$$

subject to (i) $\displaystyle \alpha_0 + \sum_{n=1}^{N} \alpha_n \ln(p_{kn}^0/C_k^0) + \sum_{m=1}^{M} \beta_m \ln u_{km}^0$

$$+ \frac{1}{2}\sum_{n=1}^{N}\sum_{n'=1}^{N} \alpha_{nn'} \left(\ln(p_{kn}^0/C_k^0) \right) \left(\ln(p_{kn'}^0/C_k^0) \right)$$

$$+ \frac{1}{2}\sum_{m=1}^{M}\sum_{m'=1}^{M} \beta_{mm'} (\ln u_{km}^0)(\ln u_{km'}^0)$$

$$+ \sum_{n=1}^{N}\sum_{m=1}^{M} \gamma_{nm} \left(\ln(p_{kn}^0/C_k^0) \right) (\ln u_{km}^0) \leqq 0, k = 1, \cdots, K^0,$$

(ii) $\displaystyle \sum_{m=1}^{M} \beta_m = 1,$

$$\sum_{m'=1}^{M} \beta_{mm'} = \sum_{m=1}^{M} \gamma_{nm} = 0, m = 1, \cdots, M, n = 1, \cdots, N,$$

(iii) $\alpha_{nn'} = \alpha_{n'n}, n = 1, \cdots, N, n' = 1, \cdots, N,$

$$\beta_{mm'} = \beta_{m'm}, m = 1, \cdots, M, m' = 1, \cdots, M.$$

The restrictions (i) construct the problem as a frontier, the restrictions (ii) impose homogeneity of outputs and (iii) imposes symmetry.

The maximization over the parameters (α, β, γ) yields the translog frontier for the data $\left((p/C)^{k,0}, u^{k,0}\right), k = 1, \cdots, K^0$. Denote the solutions by $(\alpha_0^0, \alpha_n^0, \beta_m^0, \alpha_{nn'}^0, \beta_{mm'}^0, \gamma_{nm}^0)$, then these parameter values are the coefficients that form the frontier. For observation k, two components of the indirect output based Malmquist index can now be computed. First, by inserting $\left((p/C)^{k,0}, u^{k,0}\right)$ for each $k = 1, \cdots, K^0$ into the estimated function with the "o" parameters, we

145

obtain $ID_o^0\left((p/C)^{k,0}, u^{k,0}\right)$, for each observation in period "0". Next, by inserting $\left((p/C)^{k,1}, u^{k,1}\right)$ for each observation in period "1" into the same expression we obtain $ID_o^0\left((p/C)^{k,1}, u^{k,1}\right)$, i.e., the efficiency of $k = 1, \cdots, K^1$ relative to the reference technology at period "0".

The second linear programming problem we need to solve in order to compute $M_o^1\left((p/C)^{k,1}, u^{k,1}, (p/C)^{k,0}, u^{k,0}\right)$ for each k is

$$(7.3.2) \qquad \max_{(\alpha,\beta,\gamma)} \quad \sum_{k=1}^{K^1} \left[\alpha_0 + \sum_{n=1}^{N} \alpha_n \ln(p_{kn}^1/C_k^1) + \sum_{m=1}^{M} \beta_m \ln u_{km}^1 \right.$$

$$+\frac{1}{2}\sum_{n=1}^{N}\sum_{n'=1}^{N}\alpha_{nn'}\left(\ln(p_{kn}^1/C_k^1)\right)\left(\ln(p_{kn'}^1/C_k^1)\right)$$

$$+\frac{1}{2}\sum_{m=1}^{M}\sum_{m'=1}^{M}\beta_{mm'}(\ln u_{km}^1)(\ln u_{km'}^1)$$

$$\left. +\sum_{n=1}^{N}\sum_{m=1}^{M}\gamma_{nm}\left(\ln(p_{kn}^1/C_k^1)\right)(\ln u_{km}^1)\right]$$

$$\text{subject to} \quad \text{(i)} \ \alpha_0 + \sum_{n=1}^{N} \alpha_n \ln(p_{kn}^1/C_k^1) + \sum_{m=1}^{M} \beta_m \ln u_{km}^1$$

$$+\frac{1}{2}\sum_{n=1}^{N}\sum_{n'=1}^{N}\alpha_{nn'}\left(\ln(p_{kn}^1/C_k^1)\right)\left(\ln(p_{kn'}^1/C_k^1)\right)$$

$$+\frac{1}{2}\sum_{m=1}^{M}\sum_{m'=1}^{M}\beta_{mm'}(\ln u_{km}^1)(\ln u_{km'}^1)$$

$$+\sum_{n=1}^{N}\sum_{m=1}^{M}\gamma_{nm}\left(\ln(p_{kn}^1/C_k^1)\right)(\ln u_{km}^1) \leqq 0, k = 1, \cdots, K^1,$$

$$\text{(ii)} \ \sum_{m=1}^{M} \beta_m = 1,$$

$$\sum_{m'=1}^{M} \beta_{mm'} = \sum_{m=1}^{M} \gamma_{nm} = 0, m = 1, \cdots, M, n = 1, \cdots, N,$$

$$\text{(iii)} \ \alpha_{nn'} = \alpha_{n'n}, n = 1, \cdots, N, n' = 1, \cdots, N,$$

$$\beta_{mm'} = \beta_{m'm}, m = 1, \cdots, M, m' = 1, \cdots, M.$$

where (i) - (iii) impose the appropriate restrictions to ensure appropriate value of the distance function, homogeneity in outputs and symmetry as before.

Denote the solutions by $(\alpha_0^1, \alpha_n^1, \beta_m^1, \alpha_{nn'}^1, \beta_{mm'}^1, \gamma_{nm}^1)$, then for each observation k, the last two components of the output based indirect productivity index can be calculated. First by inserting $\left((p/C)^{k,1}, u^{k,1}\right), k = 1, \cdots, K^1$ into the estimated translog frontier with the "1" parameters, we obtain $ID_o^1\left((p/C)^{k,1}, u^{k,1}\right)$; for each observation in period "1". Then by inserting $\left((p/C)^{k,0}, u^{k,0}\right)$

into the same translog frontier, the last component $ID_o^1\left((p/C)^{k,0}, u^{k,0}\right)$ is determined for each observation. The four components can be substituted into (7.1.2) or (7.1.3) to yield the productivity index $M_o^1\left((p/C)^{k,1}, u^{k,1}, (p/C)^{k,0}, u^{k,0}\right)$ for $k = 1, \cdots, K$.

The direct output based Malmquist productivity index may also be computed using parametric formulations of the corresponding distance functions. Suppose that there are $k = 1, \cdots, K^t, t = 0, 1$, observations of (positive) input vectors $x^{k,t} = (x_{k1}^t, \cdots, x_{kN}^t), x_{kn}^t > 0$ and k observations of (positive) output vectors $u^{k,t} = (u_{k1}^t, \cdots, u_{kM}^t), u_{km}^t > 0$. Positivity is required because we specify a translog form. The parameters of the direct output distance function at "o" may be obtained as the solution to the linear programming problem

$$(7.3.3) \quad \max_{(\alpha,\beta,\gamma)} \quad \sum_{k=1}^{K^0} \left[\alpha_0 + \sum_{n=1}^{N} \alpha_n \ln x_{kn}^0 + \sum_{m=1}^{M} \beta_m \ln u_{km}^0 \right.$$

$$+ \frac{1}{2} \sum_{n=1}^{N} \sum_{n'=1}^{N} \alpha_{nn'} (\ln x_{kn}^0)(\ln x_{kn'}^0)$$

$$+ \frac{1}{2} \sum_{m=1}^{M} \sum_{m'=1}^{M} \beta_{mm'} (\ln u_{km}^0)(\ln u_{km'}^0)$$

$$\left. + \sum_{n=1}^{N} \sum_{m=1}^{M} \gamma_{nm} (\ln x_{kn}^0)(\ln u_{km}^0) \right]$$

$$\text{subject to} \quad (i) \quad \alpha_0 + \sum_{n=1}^{N} \alpha_n \ln x_{kn}^0 + \sum_{m=1}^{M} \beta_m \ln u_{km}^0$$

$$+ \frac{1}{2} \sum_{n=1}^{N} \sum_{n'=1}^{N} \alpha_{nn'} (\ln x_{kn}^0)(\ln x_{kn'}^0)$$

$$+ \frac{1}{2} \sum_{m=1}^{M} \sum_{m'=1}^{M} \beta_{mm'} (\ln u_{km}^0)(\ln u_{km'}^0)$$

$$+ \sum_{n=1}^{N} \sum_{m=1}^{M} \gamma_{nm} (\ln x_{kn}^0)(\ln u_{km}^0) \leqq 0, k = 1, \cdots, K^0,$$

$$(ii) \quad \sum_{m=1}^{M} \beta_m = 1,$$

$$\sum_{m'=1}^{M} \beta_{mm'} = \sum_{m=1}^{M} \gamma_{nm} = 0, m = 1, \cdots, M, n = 1, \cdots, N,$$

$$(iii) \quad \alpha_{nn'} = \alpha_{n'n}, n = 1, \cdots, N, n' = 1, \cdots, N,$$

$$\beta_{mm'} = \beta_{m'm}, m = 1, \cdots, M, m' = 1, \cdots, M.$$

The solution to this linear programming problem yields a set of parameters for the direct translog frontier technology, denoted by $(\alpha_0^0, \alpha_n^0, \beta_m^0, \alpha_{nn'}^0, \beta_{mm'}^0, \gamma_{nm}^0)$. For each observation k, we can use

these parameters to compute two components of the direct productivity index
$M_o^1(x^{k,1}, u^{k,1}, x^{k,0}, u^{k,0})$, namely $D_o^0(x^{k,0}, u^{k,0})$ and $D_o^0(x^{k,1}, u^{k,1})$. The second set of parameters
needed to compute the direct output based Malmquist productivity is obtained as the solution to
the linear programming problem

$$
(7.3.4) \quad \max_{(\alpha,\beta,\gamma)} \quad \sum_{k=1}^{K^1} \left[\alpha_0 + \sum_{n=1}^{N} \alpha_n \ln x_{kn}^1 + \sum_{m=1}^{M} \beta_m \ln u_{km}^1 \right.
$$

$$
+\frac{1}{2} \sum_{n=1}^{N} \sum_{n'=1}^{N} \alpha_{nn'}(\ln x_{kn}^1)(\ln x_{kn'}^1)
$$

$$
+\frac{1}{2} \sum_{m=1}^{M} \sum_{m'=1}^{M} \beta_{mm'}(\ln u_{km}^1)(\ln u_{km'}^1)
$$

$$
\left. \sum_{n=1}^{N} \sum_{m=1}^{M} \gamma_{nm}(\ln x_{kn}^1)(\ln u_{km}^1) \right]
$$

$$
\text{subject to} \quad \text{(i)} \; \alpha_0 + \sum_{n=1}^{N} \alpha_n \ln x_{kn}^1 + \sum_{m=1}^{M} \beta_m \ln u_{km}^1
$$

$$
+\frac{1}{2} \sum_{n=1}^{N} \sum_{n'=1}^{N} \alpha_{nn'}(\ln x_{kn}^1)(\ln x_{kn'}^1)
$$

$$
+\frac{1}{2} \sum_{m=1}^{M} \sum_{m'=1}^{M} \beta_{mm'}(\ln u_{km}^1)(\ln u_{km'}^1)
$$

$$
\sum_{n=1}^{N} \sum_{m=1}^{M} \gamma_{nm}(\ln x_{kn}^1)(\ln u_{km}^1) \leqq 0, k = 1, \cdots, K^1,
$$

$$
\text{(ii)} \; \sum_{m=1}^{M} \beta_m = 1,
$$

$$
\sum_{m'=1}^{M} \beta_{mm'} = \sum_{m=1}^{M} \gamma_{nm} = 0, m = 1, \cdots, M, n = 1, \cdots, N,
$$

$$
\text{(iii)} \; \alpha_{nn'} = \alpha_{n'n}, n = 1, \cdots, N, n' = 1, \cdots, N,
$$

$$
\beta_{mm'} = \beta_{m'm}, m = 1, \cdots, M, m' = 1, \cdots, M.
$$

The parameter values associated with the solution to (7.3.4) are denoted by
$(\alpha_0^1, \alpha_n^1, \beta_m^1, \alpha_{nn'}^1, \beta_{mm'}^1, \gamma_{nm}^1)$. These values are used to determine the last two components of the
productivity index, i.e., for each observation k, $D_o^1(x^{k,1}, u^{k,1})$ and $D_o^1(x^{k,0}, u^{k,0})$ are computed.
Thus from the two linear programming problems, (7.3.3) and (7.3.4) the direct output based
Malmquist productivity index can be computed for each observation k, by substituting into
(7.1.5).

The direct and indirect input based Malmquist productivity indexes may also be computed using specific functional forms. Here we continue the use of the translog as the specification, but of course other forms that are linear in parameters apply equally well.

Suppose that for each $t = 0, 1$, there are $k = 1, \cdots, K^t$ observations of revenue deflated output prices $(r/R)^{k,t} = (r_{k1}^t/R_k^t, \cdots, r_{kM}^t/R_k^t)$, each $(r_{km}^t/R_k^t) > 0$, and inputs $x^{k,t} = (x_{k1}^t, \cdots, x_{kN}^t)$, each $x_{kn}^t > 0$, where positivity is required because we take logs of the data. The parameters of the revenue indirect input distance function at "o" may be calculated as the solution to the linear programming problem

$$(7.3.5) \quad \min_{(\alpha,\beta,\gamma)} \sum_{k=1}^{K^0} \left[\alpha_0 + \sum_{m=1}^{M} \alpha_m \ln(r_{km}^0/R_k^0) + \sum_{n=1}^{N} \beta_n \ln x_{kn}^0 \right.$$

$$+ \frac{1}{2} \sum_{m=1}^{M} \sum_{m'=1}^{M} \alpha_{mm'} \left(\ln(r_{km}^0/R_k^0) \right) \left(\ln(r_{km'}^0/R_k^0) \right)$$

$$+ \frac{1}{2} \sum_{n=1}^{N} \sum_{n'=1}^{N} \beta_{nn'}(\ln x_{kn}^0)(\ln x_{kn'}^0)$$

$$\left. + \sum_{m=1}^{M} \sum_{n=1}^{N} \gamma_{mn} \left(\ln(r_{km}^0/R_k^0) \right) (\ln x_{kn}^0) \right]$$

subject to (i) $\alpha_0 + \sum_{m=1}^{M} \alpha_m \ln(r_{km}^0/R_k^0) + \sum_{n=1}^{N} \beta_n \ln x_{kn}^0$

$$+ \frac{1}{2} \sum_{m=1}^{M} \sum_{m'=1}^{M} \alpha_{mm'}(\ln(r_{km}^0/R_k^0))(\ln(r_{km'}^0/R_k^0))$$

$$+ \frac{1}{2} \sum_{n=1}^{N} \sum_{n'=1}^{N} \beta_{nn'}(\ln x_{kn}^0)(\ln x_{kn'}^0)$$

$$+ \sum_{m=1}^{M} \sum_{n=1}^{N} \gamma_{mn} \left(\ln(r_{km}^0/R_k^0) \right) (\ln x_{kn}^0) \geqq 0, k = 1, \cdots, K^0,$$

(ii) $\sum_{n=1}^{N} \beta_n = 1,$

$$\sum_{n'=1}^{N} \beta_{nn'} = \sum_{n=1}^{N} \gamma_{mn} = 0, n = 1, \cdots, N, m = 1, \cdots, M,$$

(iii) $\alpha_{mm'} = \alpha_{m'm}, m = 1, \cdots, M, m' = 1, \cdots, M,$

$$\beta_{nn'} = \beta_{n'n}, n = 1, \cdots, N, n' = 1, \cdots, N.$$

The restrictions (i) - (iii) ensure that the distance function takes appropriate values, that it is homogeneous in inputs and satisfies symmetry.

The solution to the above linear programming problem yields the first set of parameters namely $(\alpha_o^0, \alpha_n^0, \beta_m^0, \alpha_{nn'}^0, \beta_{mm'}^0, \gamma_{mn}^0)$. These parameters "span" the translog frontier technology at "o". Together with data on each observation k, the two components $ID_i^0\left((r/R)^{k,0}, x^{k,0}\right)$ and $ID_i^0\left((r/R)^{k,1}, x^{k,1}\right)$ of this productivity index may be calculated. The remaining two components are calculated using the coefficients from

$$
\begin{aligned}
(7.3.6) \quad \min_{(\alpha,\beta,\gamma)} \quad \sum_{k=1}^{K^1} \Bigg[&\alpha_0 + \sum_{m=1}^{M} \alpha_m \ln(r_{km}^1/R_k^1) + \sum_{n=1}^{N} \beta_n \ln x_{kn}^1 \\
&+ \frac{1}{2} \sum_{m=1}^{M} \sum_{m'=1}^{M} \alpha_{mm'} \left(\ln(r_{km}^1/R_k^1)\right)\left(\ln(r_{km'}^1/R_k^1)\right) \\
&+ \frac{1}{2} \sum_{n=1}^{N} \sum_{n'=1}^{N} \beta_{nn'} (\ln x_{kn}^1)(\ln x_{kn'}^1) \\
&+ \sum_{m=1}^{M} \sum_{n=1}^{N} \gamma_{mn} \left(\ln(r_{km}^1/R_k^1)\right)(\ln x_{kn}^1) \Bigg]
\end{aligned}
$$

subject to : (i) $\alpha_0 + \sum_{m=1}^{M} \alpha_m \ln(r_{km}^1/R_k^1) + \sum_{n=1}^{N} \beta_n \ln x_{km}^1$

$$
\frac{1}{2} \sum_{m=1}^{M} \sum_{m'=1}^{M} \alpha_{mm'} \left(\ln(r_{km}^1/R_k^1)\right)\left(\ln(r_{km'}^1/R_k^1)\right)
$$

$$
+ \frac{1}{2} \sum_{n=1}^{N} \sum_{n'=1}^{N} \beta_{nn'} (\ln x_{kn}^1)(\ln x_{kn'}^1)
$$

$$
+ \sum_{m=1}^{M} \sum_{n=1}^{N} \gamma_{mn} \left(\ln(r_{km}^1/R_k^1)\right)(\ln x_{kn}^1) \geqq 0, k = 1, \cdots, K^1,
$$

(ii) $\sum_{n=1}^{N} \beta_n = 1$,

$$
\sum_{n'=1}^{N} \beta_{nn'} = \sum_{n=1}^{N} \gamma_{mn} = 0, n = 1, \cdots, N, m = 1, \cdots, M,
$$

(iii) $\alpha_{mm'} = \alpha_{m'm}, m = 1, \cdots, M, m' = 1, \cdots, M,$

$$
\beta_{nn'} = \beta_{n'n}, n = 1, \cdots, N, n' = 1, \cdots, N.
$$

The parameters that solve the linear programming problem (7.3.6) are denoted by $(\alpha_o^1, \alpha_n^1, \beta_m^1, \alpha_{nn'}^1, \beta_{mm'}^1, \gamma_{mn}^1)$. These together with data on each observation k yield the two remaining components in the input based indirect productivity index, namely $ID_i^1\left((r/R)^{k,1}, x^{k,1}\right)$ and $ID_i^1\left((r/R)^{k,0}, x^{k,0}\right)$. The four components yield the productivity measure for each observation upon substitution into (7.1.7) or (7.1.8).

We leave to the reader the specification for the direct input based Malmquist productivity measure.

7.4 The Fisher Ideal Index Approach to Productivity Measurement

In the previous two sections linear programming techniques are used to calculate Malmquist productivity indexes. In Section 7.2 no specific functional form is imposed on the data while in Section 7.3 translog parameterizations of the distance functions were chosen. The first method may be termed nonparametric while the second method is parametric. Here a second nonparametric approach for calculation of productivity is introduced.

Denote input prices by p and suppose that for each $t, t = 0, 1$ there are $k = 1, \cdots, K^t$ observations of $p^{k,t} = (p^t_{k1}, \cdots, p^t_{kN})$. Also suppose that there are K observations of input vectors $x^{k,t} = (x^t_{k1}, \cdots, x^t_{kN})$. Recall that for k the *Paasche Input Quantity Index* is written as

$$(7.4.1) \qquad \frac{\sum_{n=1}^{N} p^1_{kn} x^1_{kn}}{\sum_{n=1}^{N} p^1_{kn} x^0_{kn}} = \frac{p^{k,1} x^{k,1}}{p^{k,1} x^{k,0}},$$

i.e., the ratio of the value of two input vectors at "0" and "1" evaluated with respect to the "1" prices. Also recall that for observation k, the *Laspeyres Input Quantity Index* is written as

$$(7.4.2) \qquad \frac{\sum_{n=1}^{N} p^0_{kn} x^1_{kn}}{\sum_{n=1}^{N} p^0_{kn} x^0_{kn}} = \frac{p^{k,0} x^{k,1}}{p^{k,0} x^{k,0}},$$

i.e., the ratio of the value of two input vectors $x^{k,0}$ and $x^{k,1}$ evaluated at the input prices prevailing in period "0". The *Fisher Ideal Input Quantity Index* is the geometric mean of the corresponding Paasche and Laspeyres indexes, namely

$$(7.4.3) \qquad FI^1_i(p^{k,1}, x^{k,1}, p^{k,0}, x^{k,0}) = \left[\frac{p^{k,1} x^{k,1}}{p^{k,1} x^{k,0}} \frac{p^{k,0} x^{k,1}}{p^{k,0} x^{k,0}} \right]^{\frac{1}{2}}$$

Our first task is to show that under certain conditions the Fisher ideal index (7.4.3) can be shown to be equal to the direct Malmquist input based productivity index (7.1.10). Thus recall that the cost function is defined as

$$(7.4.4) \qquad C^t(u^{k,t}, p^{k,t}) = \inf_{x^t} \{ p^{k,t} x^t : x^t \in L^t(u^{k,t}) \}.$$

It now follows that

$$(7.4.5) \qquad C^t(u^{k,t}, p^{k,t}) \leq p^{k,t} x^t \text{ for all } x^t \in L^t(u^{k,t}) \text{ and } p^{k,t} \in \Re^N.$$

Since $\left(x^{k,t} / D^t_i(u^{k,t}, x^{k,t}) \right) \in L^t(u^{k,t})$,

$$(7.4.6) \qquad C^t(u^{k,t}, p^{k,t}) \leq \frac{p^{k,t} x^{k,t}}{D^t_i(u^{k,t}, x^{k,t})}, \text{ for all } p^{k,t} \in \Re^N.$$

At this stage introduce the assumption

(7.4.7)   **Assumption:** $L^t(u)$ is convex, $u \in \Re^M_+$.

Since $L^t(u)$ is convex, and from the maintained axioms $L^t(u)$ is closed, i.e., $(x^{k,t}/D_i^t(u^{k,t}, x^{k,t})$ belongs to the boundary of $L^t(u^{k,t})$, by the supporting hyperplane theorem there exists a price vector $p^{*,t}(u^{k,t}, x^{k,t}) = p^{*,t}, p^{*,t} \neq 0$, such that

$$(7.4.8) \qquad C^t(u^{k,t}p^{*,t}) = \frac{p^{*,t}x^{k,t}}{D_i^t(u^{k,t}, x^{k,t})}.$$

Note that if inputs are assumed to be strongly disposable, see P.2.S, then $p^{*,t} \geqq 0$. Some comments on expression (7.4.8) are in order. First, if $(p^{*,t}x^{k,t})$ equals observed cost, then in efficiency terminology, (7.4.8) expresses the condition that (direct input based) overall efficiency equals technical efficiency, i.e.,

$$(7.4.9) \qquad \frac{C^t(u^{k,t}, p^{*,t})}{p^{*,t}x^{k,t}} = 1 \Big/ D_i^t(u^{k,t}, x^{k,t}).$$

For $t = 0, 1$, we may substitute (7.4.9) into the expression for the direct input based Malmquist productivity index, where that index is defined as

$$(7.4.10) \qquad M_i^1(u^{k,1}, x^{k,1}, u^{k,0}, x^{k,0}) = \left[\frac{D_i^0(u^{k,1}, x^{k,1})}{D_i^0(u^{k,0}, x^{k,0})} \frac{D_i^1(u^{k,1}, x^{k,1})}{D_i^1(u^{k,0}, x^{k,0})} \right]^{\frac{1}{2}}.$$

In order to complete the index, observation $(x^{k,0}, u^{k,0})$ must be related to the technology at "1" and $(x^{k,1}, u^{k,1})$ must also be related to the technology of the other period. We suppose that $x^{k,0}$ can be radially expanded/contracted so that $\left(x^{k,0} \big/ D_i^1(u^{k,0}, x^{k,0}) \right)$ belongs to the isoquant of $L^1(u^{k,0})$. Under this assumption as well as convexity, the above argument may be repeated to obtain

$$(7.4.11) \qquad C^1(u^{k,0}, p^{\hat{\ast},1}) = \frac{p^{\hat{\ast},1}x^{k,0}}{D_i^1(u^{k,0}, x^{k,0})}$$

and consequently

$$(7.4.12) \qquad \frac{C^1(u^{k,0}, p^{\hat{\ast},1})}{p^{\hat{\ast},1}x^{k,0}} = 1 \Big/ D_i^1(u^{k,0}, x^{k,0}).$$

Also under similar assumptions, convexity of $L^0(u^{k,1})$ and $\left(x^{k,1} \big/ D_i^0(u^{k,1}, x^{k,1}) \right) \in IsoqL^0(u^{k,1})$, we can show that

$$(7.4.13) \qquad \frac{C^0(u^{k,1}, p^{\hat{\ast},1}p^{\hat{\ast},0})}{p^{\hat{\ast},0}x^{k,1}} = 1 \Big/ D_i^0(u^{k,1}, x^{k,1}).$$

Inserting (7.4.9) for $t = 0,1$ into (7.4.10) together with (7.4.12) and (7.4.13) yields

$$(7.4.14) \quad \left[\frac{p^{*,1}x^{k,1}}{p^{\tilde{*},1}x^{k,0}} \frac{p^{\tilde{*},0}x^{k,1}}{p^{*,0}x^{k,0}}\right]^{\frac{1}{2}} \left[\frac{C^0(u^{k,0},p^{*,0})}{C^0(u^{k,1},p^{\tilde{*},0})} \frac{C^1(u^{k,0},p^{\tilde{*},1})}{C^1(u^{k,1},p^{*,1})}\right]^{\frac{1}{2}}.$$

It is important to note that $p^{*,t}$, and $p^{\tilde{*},t}$, $t = 0,1$ depend on $(x^{k,t}, u^{k,t})$. Given this remark, introduce

(7.4.15)   **Assumption**: The optimal prices $p^{*,t}$ and $p^{\tilde{*},t}$ equal observed prices $p^{k,t}, t = 0,1$.

Assumption (7.4.15) imposes allocative efficiency on the observations, but not overall efficiency, i.e., the input vectors may be technically inefficient, but they must be allocatively efficient relative to $p^{*,t}$ and $p^{\tilde{*},t}$. Whenever Assumption (7.4.15) is valid, and output is unchanged between periods, i.e., $u^{k,0} = u^{k,1}$, we have

$$(7.4.16) \quad FI_i^1(p^{k,1}, x^{k,1}, p^{k,0}, x^{k,0}) = M_i^1(u^{k,1}, x^{k,1}, u^{k,0}, x^{k,0}),$$

or equivalently, the Fisher ideal input quantity index and the (direct) input based Malmquist productivity index yield the same outcome. However, the requirement that outputs do not change between periods is clearly unwanted. Therefore, consider the second part of (7.4.14) and in particular the first order condition of the expression

$$\max_u r^{k,t}u - C^t(u, p^{k,t}),$$

namely

$$r_m^{k,t} = \frac{\partial C^t(u, p^{k,t})}{\partial u_m}.$$

Multiply each side by $u_m^{k,t}/C^t(u^{k,t}, p^{k,t})$ and sum over m, then

$$(7.4.17) \quad r^{k,t}u^{k,t} = \varepsilon_C \cdot C^t(u^{k,t}, p^{k,t}),$$

where ε_C denotes scale elasticity, see (5.3.7). If we assume that the technology satisfies constant returns to scale, then we have $\varepsilon_C = 1$. Moreover, if we repeat this procedure for each cost function in (7.4.14) we obtain

$$(7.4.18) \quad M_i^1(u^{k,1}, x^{k,1}, u^{k,0}, x^{k,0}) = FI_i^1(p^{k,1}, x^{k,1}, p^{k,0}, x^{k,0})\Big/FI_o^1(r^{k,1}, u^{k,1}, r^{k,0}, u^{k,0}),$$

153

where F_o^1 denotes the Fisher ideal output quantity index, see (7.4.31). Thus under constant return to scale and no allocative inefficiency, the Malmquist input based productivity index is the quotient of the Fisher input and output quantity indexes.

Next we consider the indirect input based Malmquist productivity index. For observation k it may be written as

$$(7.4.19) \quad IM_i^1\left((r/R)^{k,1}, x^{k,1}, (r/R)^{k,0}, x^{k,0}\right) = \left[\frac{ID_i^0\left((r/R)^{k,1}, x^{k,1}\right)}{ID_i^0\left((r/R)^{k,0}, x^{k,0}\right)} \frac{ID_i^1\left((r/R)^{k,1}x^{k,1}\right)}{ID_i^1\left((r/R)^{k,0}, x^{k,0}\right)}\right]^{\frac{1}{2}}.$$

Of the four components, let us first analyze the two for which the observation is compared to the technology in its own period, i.e., $ID_i^t\left((r/R)^{k,t}, x^{k,t}\right)$, $t = 0, 1$. Recall from Section 4.2 that the revenue indirect cost function is defined as

$$(7.4.20) \quad IC^t\left((r/R)^{k,t}, p^{k,t}\right) = \inf_{x^t}\left\{p^{k,t}x^t : x^t \in IL^t\left((r/R)^{k,t}\right)\right\}.$$

It follows that

$$(7.4.21) \quad IC^t\left((r/R)^{k,t}, p^{k,t}\right) \leqq p^{k,t}x^t \text{ for all } x^t \in IL^t\left((r/R)^{k,t}\right) \text{ and } p^{k,t} \in \Re^N.$$

Since $\left(x^{k,t}\big/ID_i^t\left((r/R)^{k,t}, x^{k,t}\right)\right) \in IL^t\left((r/R)^{k,t}\right)$,

$$(7.4.22) \quad IC^t\left((r/R)^{k,t}, p^{k,t}\right) \leqq \frac{p^{k,t}, x^{k,t}}{ID_i^t((r/R)^{k,t}, x^{k,t})} \text{ for all } p^{k,t} \in \Re^N.$$

If we assume that $IL^t\left((r/R)^{k,t}\right)$ is convex in addition to the maintained axioms, then since $\left(x^{k,t}\big/ID_i^t((r/R)^{k,t}, x^{k,t})\right)$ belongs to the isoquant of $IL^t\left((r/R)^{k,t}\right)$, there exists a price vector $p^{*,t}\left((r/R)^{k,t}, x^{k,t}\right) = p^{*,t}$, with $p^{*,t} \neq 0$, such that

$$(7.4.23) \quad IC^t\left((r/R)^{k,t}, p^{*,t}\right) = \frac{p^{*,t}x^{k,t}}{ID_i^t((r/R)^{k,t}, x^{k,t})}.$$

This expression implies that

$$(7.4.24) \quad \frac{IC^t((r/R)^{k,t}, p^{*,t})}{p^{*,t}x^{k,t}} = 1\big/ID_i^t\left((r/R)^{k,t}, x^{k,t}\right).$$

Hence, two components in the indirect Malmquist index (7.4.19) are determined. In the same way we can show that

$$(7.4.25) \quad \frac{IC^1((r/R)^{k,0}, p^{*,1})}{p^{*,1}x^{k,0}} = 1\big/ID_i^1\left((r/R)^{k,0}, x^{k,0}\right)$$

154

(7.4.26) $\quad \dfrac{IC^0((r/R)^{k,1}, p^{\hat{*},0})}{p^{\hat{*},0} x^{k,1}} = 1 \Big/ ID_i^0 \left((r/R)^{k,1}, x^{k,1}\right)$

provided that $\left(x^{k,0} \big/ ID_i^1 \left((r/R)^{k,0}, x^{k,0}\right)\right)$ belongs to the isoquant of $IL^1 \left((r/R)^{k,0}\right)$ and $\left(x^{k,1} \big/ ID_i^0 \left((r/R)^{k,1}, x^{k,1}\right)\right)$ belongs to the isoquant of $IL^0 \left((r/R)^{k,1}\right)$. The last three expressions inserted into (7.4.19) yield

(7.4.27) $\quad \left[\dfrac{p^{*,1} x^{k,1}}{p^{\hat{*},1} x^{k,0}} \dfrac{p^{\hat{*},0} x^{k,1}}{p^{*,0} x^{k,0}}\right]^{\frac{1}{2}} \left[\dfrac{IC^0((r/R)^{k,0}, p^{*,0})}{IC^0((r/R)^{k,1}, p^{\hat{*},0})} \dfrac{IC^1((r/R)^{k,0}, p^{\hat{*},1})}{IC^1((r/R)^{k,1}, p^{*,1})}\right]^{\frac{1}{2}} .$

The last expression is similar to (7.4.14), however, in (7.4.14) the prices depend on $(x^{k,t}, u^{k,t})$ while here they depend on $(x^{k,t}, (r/R)^{k,t})$. If Assumption (7.4.15) applies and revenue deflated output prices are the same in both periods, then it follows that

(7.4.28) $\quad FI_i^1(p^{k,1}, x^{k,1}, p^{k,0}, x^{k,0}) = IM_i^1 \left((r/R)^{k,1}, x^{k,1}, (r/R)^{k,0}, x^{k,0}\right),$

i.e., the Fisher ideal input quantity index and the indirect input based Malmquist productivity index yield the same outcome. Moreover, (7.4.28) together with (7.4.16) show that the direct and indirect Malmquist indexes also yield the same outcome. Of course, in the direct case prices depend on input and output quantities while in the indirect case they depend on input quantities and revenue deflated output prices.

Turning to the output based measures, as above, we first introduce the corresponding Fisher ideal index. Denote output prices by r and suppose that at each $t = 0, 1$, there are $k = 1, \cdots, K^t$ observations $r^{k,t} = (r_{k1}^t, \cdots, r_{kM}^t)$, also suppose there are k observations of output vectors $u^{k,t} = (u_{k1}^t, \cdots, u_{kM}^t)$. For each observation k, the *Paasche Output Quantity Index* is written as

(7.4.29) $\quad \dfrac{\sum_{m=1}^{M} r_{km}^1 u_{km}^1}{\sum_{m=1}^{M} r_{km}^1 u_{km}^0} = \dfrac{r^{k,1} u^{k,1}}{r^{k,1} u^{k,0}},$

i.e., the ratio of the value of two output vectors or revenue "0" and "1" evaluated at the "1" prices. Also recall that for each observation k the *Laspeyres Output Quantity Index* is written as

(7.4.30) $\quad \dfrac{\sum_{m=1}^{M} r_{km}^0 u_{km}^1}{\sum_{m=1}^{M} r_{km}^0 u_{km}^0} = \dfrac{r^{k,0} u^{k,1}}{r^{k,0} u^{k,0}},$

i.e., the ratio of the value (revenue) of the two output vectors $u^{k,0}$ and $u^{k,1}$ evaluated at the output prices at "0". The *Fisher Ideal Output Quantity Index* is the geometric mean of the Paasche and Laspeyres quantity indexes, i.e.,

(7.4.31) $\quad FI_o^1(r^{k,1}, u^{k,1}, r^{k,0}, u^{k,0}) = \left[\dfrac{r^{k,1} u^{k,1}}{r^{k,1} u^{k,0}} \dfrac{r^{k,0} u^{k,1}}{r^{k,0} u^{k,0}}\right]^{\frac{1}{2}} .$

155

To show when the Fisher output index may equal the direct Malmquist output index (7.1.5), we recall that the revenue function is defined as

(7.4.32) $R^t(x^{k,t}, r^{k,t}) = \max_{u^t} \{ r^{k,t} u^t : u^t \in P^t(x^{k,t}) \}.$

It follows that

(7.4.33) $R^t(x^{k,t}, r^{k,t}) \geq r^{k,t} u^t$, for all $u^t \in P^t(x^{k,t})$ and $r^{k,t} \in \Re^M$.

Next, since $\left(u^{k,t} / D_o^t(x^{k,t}, u^{k,t}) \right) \in P^t(x^{k,t})$,

(7.4.34) $R^t(x^{k,t}, r^{k,t}) \geqq \dfrac{r^{k,t} u^{k,t}}{D_o^t(x^{k,t}, u^{k,t})}$, for all $r^{k,t} \in \Re^M$.

Next, if we assume that $P^t(x^{k,t})$ is convex, then by the maintained axioms $(u^{k,t} / D_o^t(x^{k,t}, u^{k,t})) \in Isoq P^t(x^{k,t})$, and there exists an output price vector $r^{*,t}(x^{k,t}, u^{k,t}) = r^{*,t}, r^{*,t} \neq 0$, such that

(7.4.35) $R^t(x^{k,t}, r^{*,t}) = \dfrac{r^{*,t} u^{k,t}}{D_o^t(x^{k,t}, u^{k,t})}.$

If outputs are strongly disposable, see P.3.S, then $r^{*,t} \geqq 0$, however here we only impose weak disposability, P.3, thus some prices, but not all, can be negative. (7.4.35) gives us two terms that can be substituted into (7.1.5), i.e.,

(7.4.36) $\dfrac{R^t(x^{k,t}, r^{*,t})}{r^{*,t} u^{k,t}} = 1 \Big/ D_o^t(x^{k,t}, u^{k,t}), t = 0, 1.$

The two remaining components can be derived in the same manner as (7.4.36), however since the reference technologies and observations in these two cases are of different periods, one must gaurantee that $(u^{k,1} / D_o^0(x^{k,1}, u^{k,1})) \in P^0(x^{k,1})$ and that $(u^{k,0} / D_o^1(x^{k,0}, u^{k,0})) \in P^1(x^{k,0})$. Whenever this holds we can prove that

(7.4.37) $\dfrac{R^1(x^{k,0}, r^{\hat{},1})}{r^{\hat{},1} u^{k,0}} = 1 \Big/ D_o^1(x^{k,0}, u^{k,0})$

(7.4.38) $\dfrac{R^0(x^{k,1}, r^{\hat{},0})}{r^{\hat{},0} u^{k,1}} = 1 \Big/ D_o^0(x^{k,1}, u^{k,1}).$

Inserting (7.4.36) for $t = 0, 1$ and (7.4.37), (7.4.38) into (7.1.5) yields

(7.4.39) $\left[\dfrac{r^{*,1} u^{k,1}}{r^{\hat{},1} u^{k,0}} \dfrac{r^{\hat{},0} u^{k,1}}{r^{*,0} u^{k,0}} \right]^{\frac{1}{2}} \left[\dfrac{R^0(x^{k,0}, r^{*,0})}{R^0(x^{k,1}, r^{\hat{},0})} \dfrac{R^1(x^{k,0}, r^{\hat{},1})}{R^1(x^{k,1}, r^{*,1})} \right]^{\frac{1}{2}}.$

It is important to keep in mind that the price vectors $r^{*,t}$, and $r^{\hat{},t}$, $t = 0, 1$, are functions of $x^{k,t}$ and $u^{k,t}$.

To complete our task, introduce

(7.4.40)   **Assumption:** The optimal prices $r^{*,t}$ and $r^{\hat{*},t}$ equal observed prices $r^{k,t}, t = 0, 1$.

This assumption imposes allocative but not overall efficiency on the observation, i.e., an observation can be technically inefficient. Whenever Assumption (7.4.40) is valid and $x^{k,0} = x^{k,1}$, it follows that

$$(7.4.41) \quad FI_o^1(r^{k,1}, u^{k,1}, r^{k,0}, u^{k,0}) = M_o^1(x^1, u^1, x^0, u^0),$$

i.e., the Fisher ideal output quantity index equals the direct output based Malmquist index.

Next we turn our attention to the indirect output based Malmquist productivity index. For observation k it can be written as

$$(7.4.42) \quad IM_o^1\left((p/C)^{k,1}, u^{k,1}, (p/C)^{k,0}, u^{k,0}\right) = \left[\frac{ID_o^0\left((p/C)^{k,1}, u^{k,1}\right)}{ID_o^0\left((p/C)^{k,0}, u^{k,0}\right)} \frac{ID_o^1\left((p/C)^{k,1}, u^{k,1}\right)}{ID_o^1\left((p/C)^{k,0}, u^{k,0}\right)}\right]^{\frac{1}{2}}.$$

To show that this index may equal the output based Fisher ideal index, recall from Section 4.1 that the indirect revenue function is defined as

$$(7.4.43) \quad IR^t\left((p/C)^{k,t}, r^{k,t}\right) = \max_{u^t}\left\{r^{k,t}u^t : u^t \in IP^t\left((p/C)^{k,t}\right)\right\}.$$

If we assume that $IP^t\left((p/C)^{k,t}\right)$ is convex, then using (7.4.42) we can prove that there exist $r^{*,t}$ such that

$$(7.4.44) \quad \frac{IR^t((p/C)^{k,t}, r^{*,t})}{r^{*,t}u^{k,t}} = 1 \Big/ ID_o^t\left((p/C)^{k,t}, u^{k,t}\right).$$

Moreover, under certain conditions (what are these?) one may prove that

$$(7.4.45) \quad \frac{IR^1((p/C)^{k,0}, r^{\hat{*},1})}{r^{\hat{*},1}u^{k,0}} = 1 \Big/ ID_o^1\left((p/C)^{k,0}, u^{k,0}\right)$$

$$(7.4.46) \quad \frac{IR^0((p/C)^{k,0}, r^{\hat{*},0})}{r^{\hat{*},0}u^{k,1}} = 1 \Big/ ID_o^0\left((p/C)^{k,1}, u^{k,1}\right).$$

Inserting the last three expressions into (7.4.41) yields

$$(7.4.47) \quad \left[\frac{r^{*,1}u^{k,1}}{r^{*,1}u^{k,0}}\frac{r^{*,0}u^{k,1}}{r^{*,0}u^{k,0}}\right]^{\frac{1}{2}} \cdot \left[\frac{IR^0((p/C)^{k,0}, r^{*,0})}{IR^0((p/R)^{k,1}, r^{\hat{*},0})}\frac{IR^1((p/C)^{k,0}, r^{\hat{*},1})}{IR^1((p/C)^{k,1}, r^{*,1})}\right]^{\frac{1}{2}}.$$

If Assumption (7.4.39) applies, and $(p/C)^{k,0} = (p/C)^{k,1}$ then it follows that

$$(7.4.48) \quad FI_o^1(r^{k,1}, u^{k,1}, r^{k,0}, u^{k,0}) = IM_o^1\left((p/C)^{k,1}, u^{k,1}, (p/C)^{k,0}, u^{k,0}\right).$$

This shows that under certain conditions the Fisher ideal output quantity index and the indirect output based Malmquist productivity index are equal. Note also that this implies that the indirect and direct Malmquist indexes coincide, see (7.4.41).

7.5 The Törnqvist Index Approach to Productivity Measurement

In the two earlier sections, 7.2 and 7.4, nonparametric techniques for computing productivity were discussed. A third nonparametric method for gauging productivity due to Caves, Christensen and Diewert (1982) is developed in this section. The main tools are translog distance functions and Törnqvist indexes (to be defined). The general approach employed in this section is to parameterize distance functions as translog, and, by invoking the translog identity (see Appendix), show that these can be written as Törnqvist indexes.

To be specific, suppose there are $k = 1, \cdots, K^t, t = 0, 1$ observations on (positive) input quantities $x^{k,t} = (x_{k1}^t, \cdots, x_{kN}^t)$ and input prices $p^{k,t} = (p_{k1}^t, \cdots, p_{kN}^t)$. The *Törnqvist Input Quantity Index* is defined for k as

$$(7.5.1) \qquad \ln T_i^1(p^{k,1}, x^{k,1}, p^{k,0}, x^{k,0}) = \frac{1}{2} \left[\sum_{n=1}^N \left(\frac{p_{kn}^1 x_{kn}^1}{p^{k,1} x^{k,1}} + \frac{p_{kn}^0 x_{kn}^0}{p^{k,0} x^{k,0}} \right) \left(\ln x_{kn}^1 - \ln x_{kn}^0 \right) \right]$$

If we denote the shares $\frac{p_{kn}^t x_{kn}^t}{p^{k,t} x^{k,t}}$ as $s_n^{k,t}$, then we may also write (7.5.1) as

$$(7.5.2) \qquad T_i^1(p^{k,1}, x^{k,1}, p^{k,0}, x^{k,0}) = \prod_{n=1}^N \left(\frac{x_{kn}^1}{x_{kn}^0} \right)^{\sum_{n=1}^N \left(s_n^{k,1} + s_n^{k,0} \right)}$$

In the same spirit as (7.5.1) we define the *Törnqvist Output Quantity Index* for k as

$$(7.5.3) \qquad \ln T_o^1(r^{k,0}, u^{k,1}, r^{k,0}, u^{k,0}) = \frac{1}{2} \left[\sum_{m=1}^M \left(\frac{r_{km}^1 u_{km}^1}{r^{k,1} u^{k,1}} + \frac{r_{km}^0 u_{km}^0}{r^{k,0} u^{k,0}} \right) \left(\ln u_{km}^1 - \ln u_{km}^0 \right) \right],$$

where $r^{k,t} = (r_{k1}^t, \cdots, r_{kM}^t), t = 0, 1$, denotes observed output prices and $u^{k,t} = (u_{k1}^t, \cdots, u_{kM}^t), t = 0, 1$, denotes (positive) observed output quantities.

To show how the above two Törnqvist index are used in modeling productivity, suppose that the distance functions in the Malmquist productivity index (7.1.5) are translog output distance functions, as in Section 7.3, specifically for each k, and $t = 0, 1$,

$$\begin{aligned}
(7.5.4) \qquad \ln D_o^t(x^{k,t}, u^{k,t}) =\ & \alpha_o^t + \sum_{n=1}^N \alpha_n^t \ln x_{kn}^t + \sum_{m=1}^M \beta_m^t \ln u_{km}^t \\
& + \frac{1}{2} \sum_{n=1}^N \sum_{n'=1}^N \alpha_{nn'}^t (\ln x_{kn}^t)(\ln x_{kn'}^t) \\
& + \frac{1}{2} \sum_{m=1}^M \sum_{m'=1}^M \beta_{mm'}^t (\ln u_{km}^t)(\ln u_{km'}^t) \\
& + \sum_{n=1}^N \sum_{m=1}^M \gamma_{nm}^t (\ln x_{kn}^t)(\ln u_{km}^t).
\end{aligned}$$

Now if we assume that $\alpha^0_{nn'} = \alpha^1_{nn'}, \beta^0_{mm'} = \beta^1_{mm'}, \gamma^0_{nm} = \gamma^1_{nm}, n, n' = 1, \cdots, N, m, m' = 1, \cdots, M$, then it follows from the translog identity, see Appendix, that

$$
(7.5.5) \quad \ln M^1_o(x^{k,1}, u^{k,1}, x^{k,0}, u^{k,0}) = \frac{1}{2} \left[\nabla_{\ln x} \ln D^1_o(x^{k,1}, u^{k,1}) \right.
$$
$$
\left. + \nabla_{\ln x} \ln D^0_o(x^{k,0}, u^{k,0}) \right] (\ln x^{k,1} - \ln x^{k,0})
$$
$$
+ \frac{1}{2} \left[\nabla_{\ln u} \ln D^1_o(x^{k,1}, u^{k,0}) \right.
$$
$$
\left. + \nabla_{\ln u} \ln D^0_o(x^{k,0}, u^{k,0}) \right] (\ln u^{k,1} - \ln u^{k,0}),
$$

where $\ln M^1_o(x^{k,1}, u^{k,1}, x^{k,0}, u^{k,0})$ is a Malmquist productivity index. Next, from expression
(5.2.31) it follows that $\frac{\partial \ln D^1_o(x^{k,1}, u^{k,1})}{\partial x_n} = \frac{p^1_n x^1_n}{p^1 x^{k,1}} \varepsilon_{D_o}$, where $p^1 x^{k,1}$ minimizes costs and ε_{D_o} expresses
scale elasticity. For the other inputs similar expressions can be derived. Thus turning to
$\frac{\partial \ln D^1_o(x^{k,1}, u^{k,1})}{\partial u_m}$, this is equal to $\frac{\hat{r}^1_m u^1_{km}}{D^1_o(x^{k,1}, u^{k,1})}$, where $\hat{r}^1_m$ is revenue deflated output price. Note that
$\hat{r}^1_m = r^1_m u^1_{km} / r^1 u^{k,1}$ and that it is assumed that $D^1_o(x^{k,1}, u^{k,1}) = 1$. Therefore, if observed inputs
and outputs are optimizers, then (7.5.5) becomes

$$
(7.5.6) \quad \ln M^1_o(x^{k,1}, u^{k,1}, x^{k,0}, u^{k,0}) = \frac{1}{2} \sum_{n=1}^N \left(\frac{p^1_n x^1_{kn}}{p^1 x^{k,1}} \varepsilon^1_{D_o} + \frac{p^0_n x^0_{kn}}{p^0 x^{k,0}} \varepsilon^0_{D_o} \right) (\ln x^1_{kn} - \ln x^0_{kn})
$$
$$
+ \frac{1}{2} \sum_{m=1}^M \left(\frac{r^1_m u^1_{km}}{r^1 u^{k,1}} + \frac{r^0_m u^0_{km}}{r^0 u^{k,0}} \right) (\ln u^1_{km} - \ln u^0_{km}).
$$

Thus if the technology exhibits constant returns to scale at each $t = 0, 1$ (i.e., $\varepsilon^0_{D_o} = \varepsilon^1_{D_o} = 1$),
then the (direct) output based Malmquist productivity index equals the product of the Törnqvist
input and output quantity indexes. Since the two Törnqvist indexes do not require any
estimation of functional parameters our third nonparametric method for gauging productivity has
been established.

To continue we define the *Törnqvist Cost Deflated Input Price Index* for observation k as

$$
(7.5.7) \quad \ln T^1_{(p/C)} \left((p/C)^{k,1}, x^{k,1}, (p/C)^{k,0}, x^{k,0} \right) = \frac{1}{2} \left[\sum_{n=1}^N \left(\frac{p^1_{kn} x^1_{kn}}{p^{k,1} x^{k,1}} \right. \right.
$$
$$
\left. \left. + \frac{p^0_{kn} x^0_{kn}}{p^{k,0} x^{k,0}} \right) \right] \left(\ln(p_{kn}/C^k)^1 - \ln(p_{kn}/C^k)^0 \right),
$$

where $(p/C)^{k,t}$ are cost deflated input prices and $x^{k,t}$ denotes input quantity.

Suppose that the distance functions which define the cost indirect Malmquist productivity
index $IM^1_o \left((p/C)^{k,1}, u^{k,1}, (p/C)^{k,0}, u^{k,0} \right)$ all are of the translog form with identical second order

159

terms as in Section 7.3. Then from the translog identity, see Appendix, it follows that the cost indirect Malmquist Index takes the form

$$
\begin{aligned}
\ln IM_o^1\left((p/C)^{k,1}, u^{k,1}, (p/C)^{k,0}, u^{k,0}\right) &= \tfrac{1}{2}\left[\nabla_{\ln(p/C)}\ln ID_o^1\left((p/C)^{k,1}, u^{k,1}\right)\right.\\
&\quad \left.+\nabla_{\ln(p/C)}\ln ID_o^0\left((p/C)^{k,0}, u^{k,0}\right)\right]\\
&\quad \left(\ln(p/C)^{k,1}-\ln(p/C)^{k,0}\right)\\
&\quad +\tfrac{1}{2}\left[\nabla_{\ln u}\ln ID_o^1\left((p/C)^{k,1}, u^{k,1}\right)\right.\\
&\quad \left.+\nabla_{\ln u}\ln ID_o^0\left((p/C)^{k,0}, u^{k,0}\right)\right]\\
&\quad (\ln u^{k,1}-\ln u^{k,0}).
\end{aligned}
$$

Now from (4.1.22) it follows that $\frac{\partial \ln ID^o((p/C)^{k,1}, u^{k,1})}{\partial (p_n/C)} = \frac{p_n^1 x_{kn}^1}{CID_o \varepsilon_c^1}$, where ε_c^1 denotes scale elasticity, see (5.3.7) and C is total cost $p^1 x^{k,1}$. Moreover, by (5.2.5) it follows that $\frac{\partial \ln ID_o((p/C)^{k,1} u^{k,1})}{\partial \ln u_m} = \frac{\hat{r}_m^1 u_{km}^1}{ID_o}$, where $\hat{r}_m^1$ is revenue deflated output price of good m at $t = 1$. These observations together with (7.5.8) show that under constant returns to scale $\varepsilon_c^1 = \varepsilon_c^0 = 1$, the cost indirect Malmquist productivity index equals the product of the two Törnqvist Indexes (7.5.3) and (7.5.7). We leave to the reader the derivation of the Törnqvist equivalences of the input based Malmquist index (7.1.10) and the indirect Malmquist index (7.1.7).

Appendix: The Translog Identity

(7.A.1) **Lemma:** Suppose the functions $G^t(v^0), G^t(v^1), t = 0, 1$, are translog functions, i.e.,

$\ln G^t(v^0) = \alpha_o^t + \sum_{s=1}^{S} \alpha_s^t \ln v_x^0 + \tfrac{1}{2}\sum_{s=1}^{S}\sum_{s'=1}^{S}\beta_{ss'}^t(\ln v_s^0)(\ln v_{s'}^0), t = 0, 1$, then if

$\beta_{ss'}^0 = \beta_{ss'}^1$ and $v_s^t > 0, t = 0, 1$, $s, s' = 1, \cdots, S, \ln\left(G^1(v^1)/G^1(v^0)\right) +$

$\ln\left(G^0(v^1)/G^0(v^0)\right) = \left(\nabla_{\ln v}\ln G^1(v^1) + \nabla_{\ln v}\ln G^0(v^0)\right)(\ln v^1 - \ln v^0).$

Proof: See Caves, Christensen and Diewert (1982).

7.P Problems

(7.P.1) Formulate a decomposition of the Malmquist index (7.1.5) that corresponds to (7.1.3).

(7.P.2) Suppose that the data for periods "o" and "1" are $x^0 = (1, 1), u^0 = 1$ and

$x^1 = (0, 1), u^1 = 1$. Show that the "mixed period" problem (7.2.12) has no solution.

Explain. (Hint, see 7.P.4).

(7.P.3)   Write out the linear programming problem for $D_o^1(x^{k,0}, u^{k,0})$.

(7.P.4)   Prove that if $x_n^{k,0} > 0, n = 1, \cdots, K$, and inputs are strongly disposable, then $x^{k,0} \big/ D_i^1(u^{k,0}, x^{k,0})$ belongs to the isoquant of $L^1(u^{k,0})$.

(7.P.5)   Derive expression (7.4.17) without imposing constant returns to scale.

(7.P.6)   Discuss the relationship between (7.4.14) and (7.4.26).

(7.P.7)   Derive a Fisher index from the second part of (7.4.38). You may assume constant returns to scale.

(7.P.8)   Derive expression (7.4.26).

(7.P.9)   Sketch the proof of (7.4.43).

(7.P.10)   Prove the translog identity for $v^t \in \Re_+^2$.

Notes

Caves, Christensen and Diewert (1982) generalize the Malmquist (1953) work on index numbers to include productivity indexes. They also show how the (direct) Törnqvist (1936) productivity index is related to the (direct) translog formulation of the Malmquist index. Here we extend their work and the work by Färe, Grosskopf and Lovell (1990).

Appendix A

Standard Notations and Mathematical Appendix

Let A and B be two sets, we mean by

$\in$	$a \in A$	a is an element in A;
$\notin$	$a \notin A$	a is not an element in A;
$\subseteq$	$A \subseteq B$	A is a subset of B;
$\emptyset$	$A = \emptyset$	A is an empty set;
$\{a \in A: *\}$		the subset of A formed by the elements satisfying property $*$;
$\cap$	$A \cap B$	$\{x : x \in A \text{ and } x \in B\}$;
$\cup$	$A \cup B$	$\{x : x \in A \text{ or } x \in B\}$;
$\backslash$	$A \backslash B$	$\{x : x \in A, x \notin B\}$;
$+$	$A + B$	$\{z : a \in A, b \in B, z = a + b\}$;
$\Re^N$		Euclidean space of dimension N;
$\geqq$		$x, y \in \Re^N, x \geqq y$ if and only if $x_n \geqq y_n, n = 1, 2, \cdots, N$;
$\geq$		$x \geq y$ if and only if $x \geqq y$ and $x \neq y$;
$>$		$x > y$ if and only if $x_n > y_n, n = 1, 2, \cdots, N$;
$\overset{*}{>}$		$x \overset{*}{>} y$ if and only if $x_n > y_n$ or $x_n = y_n = 0, n = 1, 2, \cdots, N$;
$\Re^N_+$		$\Re^N_+ = \{x : x \in \Re^N, x \geqq 0\}$;
$\Re^N_{++}$		$\Re^N_{++} = \{x : x \in \Re^N, x > 0\}$;
$\Re^N_-$		$\Re^N_- = \{x : x \in \Re^N, x \leqq 0\}$;
$\overline{\Re}_+$		$\overline{\Re}_+ = \Re_+ \cup \{+\infty\}$;
$2^{\Re^N}$		$2^{\Re^N} = \{A : A \subseteq \Re^N\}$;

$[a, b]$	$[a, b] = \{x \in \Re : a \leq x \leq b\}$;
$[a, b)$	$[a, b) = \{x \in \Re : a \leq x < b\}$;
A is convex	for all $0 \leq \lambda \leq 1, x, y \in A, \lambda x + (1 - \lambda)y \in A$;
$\exists$	there exists;
$\forall$	for all;
$\sum$	sum sign;
$\times$	product sign;
$x^\ell \longrightarrow x^\circ$	the sequence x^ℓ converges to x°;
$\longrightarrow +\infty$	tends to $+\infty$;
$\overline{A}$	closure of $\overline{A}$;
$\Longrightarrow$	$x \in A \Longrightarrow x \in B, x$ belongs to A only if x belongs to B;
$\Longleftrightarrow$	if and only if;
$\nabla_x F(x)$	gradient of $F(x)$;
s.t.	subject to;
px	$px = \sum_{n=1}^{N} p_n x_n, p$ and $x \in \Re^N$, the inner product.
A.1	A set $A \subseteq \Re^N$ is bounded $\Longleftrightarrow \sup\{\parallel x - y \parallel : x, y \in A\} < +\infty$.
A.2	A set $F \subseteq R^N$ is closed $\Longleftrightarrow \forall x^\ell \longrightarrow x^\circ, x^\ell \in F, \forall \ell, x^\circ \in F$.
A.3	A set $Q \subseteq \Re^N$ is compact $\Longleftrightarrow$ it is closed and bounded.
A.4	A set Q in a topological space is compact $\Longleftrightarrow$ every cover of Q has a finite subcover.
A.5	A function $f : \Re_+^N \longrightarrow \Re_+$ is upper semi-continuous $\Longleftrightarrow \forall x^\ell \longrightarrow x^\circ$, $\limsup_{\ell \longrightarrow +\infty} f(x^\ell) \leq f(x^\circ) \Longleftrightarrow \{x : f(x) \geq \alpha\}$ is closed, $\forall \alpha \in \Re_+ \Longleftrightarrow GR = \{(x, \alpha) : f(x) \geq \alpha, \forall \alpha \in \Re_+\}$ is closed.
A.6	f is lower semi-continuous $\Longleftrightarrow -f$ is upper semi-continuous.
A.7	f is upper semi-continuous, Q compact, $\max_{x \in Q} f(x)$ is achieved.
A.8	$f : \Re_+^N \longrightarrow \Re_+$ is quasi-concave $\Longleftrightarrow \forall \lambda, 0 \leq \lambda \leq 1, x, y \in \Re_+^N$, $f(\lambda x + (1 - \lambda)y) \geq \min\{f(x), f(y)\}$
A.9	f is quasi-convex $\Longleftrightarrow -f$ is quasi-concave.

A.10 f is quasi-concave $\iff \{x : f(x) \geqq \alpha\}$ is convex $\forall \alpha \in \Re_+$.

A.11 f is continuous $\iff f$ is upper and lower semi-continuous.

A.12 $f: \Re \longrightarrow \Re$ is strictly increasing if $x > y \implies f(x) > f(y)$.

A.13   A continuous strictly increasing function f has an inverse f^{-1}.

A.14 $f : \Re_+^N \longrightarrow \Re_+$ is lower semi-bounded $\iff \{x : f(x) \leqq \alpha\}$ is bounded for all $\alpha \in \Re_+$.

References

Afriat, S.N. (1972) "Efficiency Estimation of Production Functions," *International Economic Review*, 13:3, pp. 568-598.

Aigner, D.J. and S.F. Chu (1968) "On Estimating the Industry Production Function," *American Economic Review*, 58:4, pp. 826-829.

Blackorby, C., D. Primont and R.R. Russell (1978) *Duality Separability, and Functional Structure: Theory and Economic Applications*, North-Holland, New York.

Caves, D., L. Christensen and E. Diewert (1982) "The Economic Theory of Index Numbers and the Measurement of Input, Output and Productivity," *Econometrica*, 50, pp. 1393-1414.

Charnes, A., W.W. Cooper, L. Seiford and J. Stutz (1983) "Invariant Multiplicative Efficiency and Piecewise Cobb-Douglas Envelopments," *Socio-Economic Planning Sciences*, pp. 223-224.

Chamber, R.G. and H. Lee (1986) "Constrained Output Maximization and U.S. Agriculture," *Applied Economics*, pp. 347-357.

Debreu, G. (1951) "The Coefficient of Resource Utilization," *Econometrica*, 19, pp. 14-22.

Debreu, G. (1959) *Theory of Value*, John Wiley and Sons, New York.

Eichhorn, W. and U. Leopold (1990) "Logical Aspects Concerning Shephard's Axioms of Production Theory," in *Generalized Convexity and Fractional Programming with Economic Applications*, A. Cambini, et al., (eds.), Springer-Verlag, Heidelberg, pp. 266-275.

Färe, R. (1988) *Fundamentals of Production Theory*, Springer-Verlag, New York.

Färe, R. and S. Grosskopf (1990) "A Distance Function Approach to Price Efficiency," *Journal of Public Economics*, pp. 123-126.

Färe, R., S. Grosskopf and C.A.K. Lovell (1988) "An Indirect Efficiency Approach to the Evaluation of Producer Performance," *Journal of Public Economics*, pp. 71-89.

Färe, R., S. Grosskopf and C.A.K. Lovell (1988) "Scale Economics and Scale Efficiency,"
Zeitschrift für die gesamte Staatwissenshaft, pp. 721-729.

Färe, R., S. Grosskopf and C.A.K. Lovell (1990) "Production Frontier," Memo.

Färe, R., S. Grosskopf and C.A.K. Lovell (1990), "Indirect Productivity Measurement," Memo.

Färe, R., S. Grosskopf and D. Njinkeu (1988) "On Piecewise Reference Technologies,"
Management Science, pp. 1507-1510.

Färe, R. and D. Primont (1990) "A Distance Function Approach to Multi-Output Technologies,"
Southern Economic Journal, pp. 879-891.

Färe, R. and K. Zieschang (1990) "Determining Output Prices in a Cost-Constrained
Not-For-Profit Organization," Mimeo.

Farrell, M.J. (1957) "The Measurement of Productive Efficiency," *Journal of the Royal
Statistical Society*, Series A, General, 120, Part 3, pp. 253-281.

Fisher, I. (1922) *The Making of Index Numbers*, Houghton Mifflin, Boston.

Fukuyama, H. (1987) "Alternative Notions and Measures of Returns to Scale: A Multi-Output
Duality Approach," Ph.D. Thesis, Southern Illinois University, Carbondale.

Førsund, F. and L. Hjalmarsson (1987) *Analyses of Industrial Structure: A Putty-Clay
Approach*, Almqvist and Wicksell, Stockholm.

Fuss, M. and D. McFadden (1978) *Production Economics, A Dual Approach to Theory and
Applications* Vol. I and II, North-Holland, Amsterdam.

Garofalo, G. and D. Malhotra (1990) "The Demand for Inputs in the Traditional Manufacturing
Region," *Applied Economics*, pp. 961-972.

Kim, Y. (1988) "Analyzing the Indirect Production Function for U.S. Manufacturing," *Southern
Economic Journal*, pp. 494-504.

Koopmans, T.C. (1951) *Activity Analysis of Production and Allocation*, John Wiley and Sons,
New York.

Lovell, C.A.K. and P. Schmidt (1988) "A Comparison of Alternative Approaches to the Measurement of Productive Efficiency," in *Application of Modern Production Theory: Efficiency and Productivity*, A. Dogramaci and R. Färe (eds.), Kluwer-Nijhoff, Boston.

Malmquist, S. (1953) "Index Numbers and Indifference Surfaces," *Trabajos de Estatistica*, 4, pp. 209-242.

von Neumann, J. (1938, 1945) "A Model of General Economic Equilibrium," *Review of Economic Studies*, 13:1, pp. 1-9.

Rockafellar, R. (1970) *Convex Analysis*, Princeton University Press, Princeton.

Shephard, R.W. (1970) *Theory of Cost and Production Functions*, Princeton University Press, Princeton.

Shephard, R.W. (1974) *Indirect Production Functions*, Verlag Anton Hain, Meisenheim am Glan.

Shephard, R.W. and R. Färe (1980) *Dynamic Theory of Production Correspondence*, Oelgeschlager, Gunn and Hain, Cambridge.

Törnqvist, L. (1936) "The Bank of Finland's Consumption Price Index," *Bank of Finland Monthly Bulletin*, 10, pp. 1-8.

Teusch, W. (1983) *Aufbau und Gewinnung Shephardscher Produktionsfunktionen unter Berücksichtigung empirischer Aspecte*, Verlag Anton Hain, Königstein.

Index